Titanium-Based Alloys - Characteristics and Applications

Edited by Petrica Vizureanu and Madalina Simona Baltatu

Published in London, United Kingdom

Titanium-Based Alloys - Characteristics and Applications
http://dx.doi.org/10.5772/intechopen.1001617
Edited by Petrica Vizureanu and Madalina Simona Baltatu

Contributors
Abdul Mohshen Sharif Ullah Siddique, Amal Mohammed Abdulrahman, Anamika Vitthal Kadam, Andrei Victor Sandu, Angel Vicente Escuder, Ayesha Khan, Dragos Cristian Achitei, Dumitru Doru Burduhos-Nergis, Igor Đorđević, Ilija Nasov, Kadephi Vuyolwethu Mjali, Kenji Fuchiwaki, Madalina Simona Baltatu, Manuela Cristina Perju, Marcelin Benchea, Mariana Correa Rossi, Marijana Popović Bajić, Marija Živković, Miguel Gomez Pólo, Minja Miličić Lazić, Nada H.A. Besisa, Pedro Peñalver, Petrică Vizureanu, Rayanah Barnawi, Ruben Agustin Panadero, Slavoljub Živković, Takeaki Yajima, Takeshi Kihara, Tankiso Lawrence Ngake, Tatsuhiko Aizawa, Tomomi Shiratori, Vicente Amigó Borrás, Vojkan Lazić, Vukoman Jokanović

First published in London, United Kingdom, 2024 by IntechOpen
IntechOpen is the global imprint of INTECHOPEN LIMITED, registered in England and Wales, registration number: 11086078, 167-169 Great Portland Street, London, W1W 5PF, United Kingdom

British Library Cataloguing-in-Publication Data
A catalogue record for this book is available from the British Library

Additional hard and PDF copies can be obtained from orders@intechopen.com

Titanium-Based Alloys - Characteristics and Applications
Edited by Petrica Vizureanu and Madalina Simona Baltatu
p. cm.
Print ISBN 978-0-85466-821-2
Online ISBN 978-0-85466-820-5
eBook (PDF) ISBN 978-0-85466-822-9

For EU product safety concerns:
IN TECH d.o.o., Prolaz Marije Krucifikse Kozulić 3, 51000 Rijeka, Croatia,
info@intechopen.com or visit our website at intechopen.com.

Meet the editors

Petrică Vizureanu obtained an MSc and Ph.D. in heating equipment at the Gheorghe Asachi Technical University, Iasi, Romania, in 1992 and 1999, respectively. Dr. Vizureanu is currently a full professor and scientific supervisor in materials engineering at the same university and the appointed director of the Department of Technologies and Equipment for Materials Processing. He is an editor and guest editor for many journals and publishing houses. His research focuses on expert systems for heating system programming, computer-assisted design for heating equipment, heating equipment for materials processing, heat transfer, biomaterials, and geopolymers. He has published more than 250 papers in international journals and conference proceedings as well as over 40 books.

Mădălina Simona Bălțatu is a lecturer at Technical University "Gh. Asachi" from Iași; she defended her Ph.D. thesis in 2017. She is a young researcher with a lot of experience in the field of biomaterials (proven by articles, international books, and book chapters: https://www.afir.org.ro/msb/). Published work includes six books and four book chapters; 60 papers published in journals and at conferences, of which 40 articles are indexed in Web of Science; five patent applications and 56 awards obtained at invention salons (32 gold medals, seven silver medals, two bronze medals, 15 special awards). More information about published work: h-index 17, Citations: 781, Source: Google Scholar; h-index 15, Citations: 495, Source: Web of Science.

Contents

Preface

At the global level, there is a continuous concern for the research and development of titanium alloys due to the excellent properties they have. The discovery of titanium dates back to the eighteenth century, but its destructive form was not successfully isolated until the twentieth century. In recent decades, titanium has gradually become important and has become the mainstay of science and technology. With the growth of the aviation industry, titanium alloys have become an ideal choice for aircraft structures. Their lightweight and high-strength properties make aircraft not only more energy efficient but also safer. Aerospace applications account for 80% of titanium consumption, while 20% of the metal is used in armor, hardware, and medical and consumer products.

Titanium-Based Alloys – Characteristics and Applications is a compact book that brings together the properties and characteristics of titanium alloys from various industrial or medical applications. In addition to their benefits, this book discusses their drawbacks and future improvements.

This book is composed of eight chapters, each addressing important aspects of titanium alloys and their applications while searching for solutions to the new generation of titanium alloys that benefit industry and medical professionals.

The chapters cover a wide range of subjects, from the use of shape-memory alloys in dentistry to the mechanical and microstructural characterization of titanium alloys processed through additive techniques. Other chapters explore the diverse applications of titanium alloys in fields such as aeronautics, medical engineering, and many others.

Intended for students, engineers, and researchers worldwide exploring the various applications and characterizations of these remarkable materials, this book brings together concepts from physics, chemistry, materials science, and engineering.

Petrică Vizureanu and Mădălina Simona Bălțatu
Gheorghe Asachi Technical University of Iași,
Iasi, Romania

Section 1

Industrial Applications

Chapter 1

Perspective Chapter: Titanium – A Versatile Metal in Modern Applications

Madalina Simona Baltatu, Petrică Vizureanu, Andrei Victor Sandu, Dragos Cristian Achitei, Manuela Cristina Perju, Dumitru Doru Burduhos-Nergis and Marcelin Benchea

Abstract

Titanium, a considerable metal renowned for its exceptional properties, has found its way into numerous industrial, medical, and aerospace applications. This chapter provides an overview of titanium's unique characteristics, which include high strength-to-weight ratio, excellent corrosion resistance, and biocompatibility, making it an ideal choice for diverse engineering and medical purposes. In the aerospace industry, titanium's low density and remarkable strength make it an essential material for aircraft components, from engine components to structural parts. Its resistance to corrosion in aggressive environments also renders it invaluable for marine applications. Medical fields have accepted titanium for orthopedic implants, dental fixtures, and surgical instruments due to its biocompatibility and ability to integrate seamlessly with living tissues. In addition to its medical and aerospace applications, titanium is used in the automotive industry for lightweight components that enhance fuel efficiency and reduce emissions.

Keywords: titanium alloys, alloying elements, properties, applications, future trends

1. Introduction

In recent decades, titanium has become one of the most valued and utilized materials across various technological and medical domains, owing to its exceptional properties. Initially discovered as dioxide in the year 1788 and extracted in its pure metallic form only by 1925, titanium has traveled a significant journey to become recognized as a revolutionary metal [1–3]. This chapter seeks to delve into the complexity and versatility of titanium, an element placed in the fourth subgroup of Mendeleev's system, alongside thorium, hafnium, and zirconium, highlighting its indispensable role in contemporary innovations.

Titanium, with atomic number 22 and an atomic weight of 47.90, is distinguished by its white-silver color, remarkable tensile strength, impressive melting and boiling points, and superior mechanical and corrosion resistance properties compared to other metals. These features render titanium and its alloys as ideal materials for a broad spectrum of applications, ranging from aerospace to medical fields, where its biocompatibility is of significant importance [4–6].

The history of titanium alloys (**Figure 1**) begins within the context of titanium's discovery and its evolving use as a metal in engineering and technology. Although pure titanium was initially isolated in its metallic form in the early nineteenth century, the development of titanium alloys truly commenced in the 1950s [7], when the demand for materials with enhanced performance in advanced applications became evident [5, 8].

Initially, titanium was considered a rare metal with limited uses due to the difficulties in extraction and processing. However, the discovery of its unique properties, such as exceptional corrosion resistance, an optimal strength-to-weight ratio, and biocompatibility, sparked interest in the development of alloys to optimize and extend these characteristics for specific applications [9–11].

In the 1950s and 1960s, with the development of the aerospace industry and space race, the demand for lightweight and strong materials increased significantly. Titanium alloys met this demand, providing ideal solutions for components of aircraft, rockets, and satellites, thanks to their resistance to extremely high temperatures and corrosion. For example, Ti-6Al-4 V, a titanium alloy with aluminum and vanadium, became one of the most utilized alloys due to its optimal balance between ductility, strength, and corrosion resistance [11].

As the processing technology for titanium evolved, other alloys were developed to meet the specific needs of various industries. In medicine, titanium alloys were adapted to maximize biocompatibility and fatigue resistance, essential for orthopedic

Figure 1.
Timeline of the discovery and use of titanium.

and dental implants. In the chemical industry and marine applications, titanium alloys were optimized for maximum corrosion resistance.

Alloying elements play an important role in modifying and improving the properties of titanium alloys, allowing them to be adapted for a wide range of industrial and medical applications. By adding different elements to the chemical composition of titanium, the mechanical properties, corrosion resistance, ductility, and machinability of the alloys can be specifically adjusted. An example is the alloy Ti-6Al-4 V, valued in the aerospace and medical industry for its balance between strength and ductility. Another alloy, Ti-6Al-7Nb, is specifically used in medical implants, providing a biocompatible alternative to traditional alloys. Alongside the well-established and widely used titanium alloy systems in industry and medicine, there is ongoing research for the development of new alloy systems that offer improved properties for specific applications. Among these, also, new titanium alloy systems are under research: Ti-Mo-Si, Ti-Mo-Zr-Mn, Ti-Mo-Zr-Ta, and Ti-Mo-Zr-Ta-Si. These alloy systems demonstrate the versatility of titanium and its capacity to be customized for specific needs (**Figure 2**) [4, 12–14].

Thus, the development of titanium alloys has been strongly associated with both industry-specific requirements and advances in technology. The defining characteristic of this process has been the ongoing collaboration of scientists, engineers, and technicians who have investigated and expanded the possibilities of this adaptable metal. With new studies focusing on enhancing mechanical qualities, creating shape memory alloys, and investigating potential in cutting-edge biomedical applications and green technology, titanium alloys continue to play a vital role in innovation today [1, 2, 9, 15, 16].

The classical methods of elaboration of Ti-based alloys for aerospace, biomedical implants, and military applications are arc remelting (VAR and EBM), selective laser melting (SLM), atomization, cold hearth melting, plasma arc welding, hot isostatic pressing (HIP), rotary forging, centrifugal casting, and sheet metal forming.

This chapter will explore the evolution of titanium's application in technology and medicine, highlighting its pivotal role in engineering innovation. It will also provide an in-depth analysis of titanium's distinctive features that contribute to its high value. In addition, we will investigate titanium alloys, the manufacturing techniques that enable its full utilization, and analyze specific applications that gain from the benefits provided by this metal [17].

Hence, this chapter will not only demonstrate the significance of titanium in the contemporary world but also emphasize its untapped capacity, opening up possibilities for novel findings and uses that have the potential to revolutionize the existing technical and medical domain.

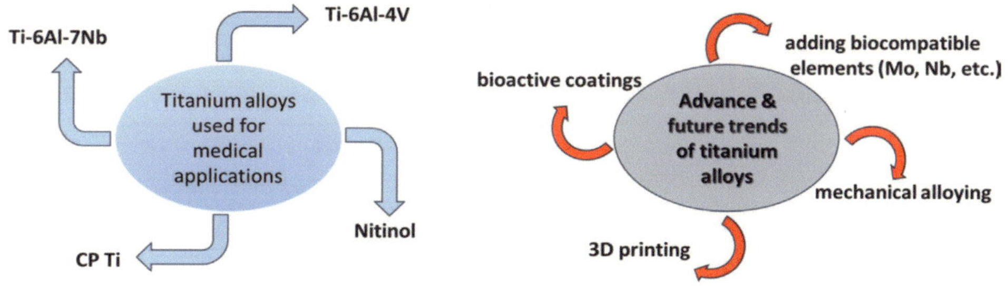

Figure 2.
Development trends of titanium alloys.

2. The unique properties of titanium

Titanium, first identified in 1791 by William Gregor, is a metallic element with a silver-gray appearance and is relatively abundant in the Earth's crust. It holds the 22nd position on the periodic table, under Group 4, and is the seventh most common element in nature (0.63%), following Al (8.8%), Fe (5.1%), Ca (3.6%), Na (2.64%), K (2.6%), and Mg (2.1%). It is part of around 100 minerals, with the most important being:

- Rutile, TiO_2
- Ilmenite, $FeTiO_3$ or $FeO.TiO_2$
- Titanomagnetite, Fe_3TiO_6 or $Fe_3O_4.TiO_2$
- Perovskite, $CaTiO_3$ or $CaO.TiO_2$
- Titanite (sphene), $CaTiSiO_5$ or $CaO.TiO_2.SiO_2$.

Ilmenite is the primary titanium mineral due to its high titanium content and ease of decomposition. Rutile, the richest titanium mineral, has been less important historically, but recent methods have made it possible to convert rutile into a very pure intermediate liquid product. Titanium production is costly due to its strong chemical bonds in existing compounds, its reactivity with many chemical elements, its absorption of gases, and the high costs associated with the metallothermic reduction of $TiCl_4$ using magnesium and sodium, as well as the purification and high-purity titanium production processes such as electrolysis or the iodide method. Despite its high cost compared to other metals, the benefits of using titanium are considerable.

Key physical properties of pure titanium are detailed in **Table 1** [18, 19].

Property	Characteristic/Value
Color in solid state	Silver-white
Density at 25°C (α-Ti)	4.51 g/cm^3
Density at 900°C (β-Ti)	4.33 g/cm^3
Melting temperature	1668°C
Thermal expansion coefficient	9.1×10^{-6}/K
Specific heat at 25°C	0.523 J/g.K
Thermal conductivity at 25°C	17–22 W/mK
Surface tension at 1600°C	1.7 N/m
Elastic modulus at 25°C	108 GN/m^2
Tensile strength	450 MPa before casting and 850 MPa after casting
Yield strength	100–200 N/m^2; 15–20%
Hardness	160–190 HB, 80–105 HV

Table 1.
Physical properties of titanium.

Titanium is a refractory metal with a melting point of 1725°C and undergoes a polymorphic transformation at 882°C, from alpha titanium with a hexagonal close-packed structure to beta titanium with a body-centered cubic structure.

When it comes to mechanical properties, titanium and its alloys have very high specific strengths. These are higher than those of steels or precipitation-hardened aluminum alloys (compare to steels in terms of strength, but with a density over 45% lower; more than twice that of aluminum alloys, with only 60% higher density). Some titanium alloys, such as Beta C, exceed, after heat treatment, a tensile strength of 1400 MPa under conditions where the density does not exceed 4.5 kg/m^3. However, very high strengths are found when alloying with a large amount of alloying elements, some of which have increased cytotoxic potential [20].

The mechanical properties of titanium distinguish it as a high-performance material, favored in many engineering and technological applications. These properties, which include tensile strength, ductility, fatigue resistance, and elastic modulus, contribute to its remarkable versatility and durability [21, 22].

The tensile strength of commercially pure titanium (with a minimum titanium content of 99%), which is characterized by excellent biocompatibility, does not exceed 550 MPa, even when it contains dissolved oxygen and nitrogen at the upper limit of the standardized range. For higher purity levels (over 99.5% Ti), the strength does not surpass 330 MPa. The hardness of pure commercial titanium does not exceed 265 HB (120 HB for grade 1 purity), and even high-strength alloys do not exceed 400 HB in the annealed state. This represents a disadvantage of titanium, which does not perform well in wear situations. Indeed, in the case of large joint prostheses (knee or hip), wear particles are responsible for many post-operative complications that can lead to implant failure.

The ductility of commercially pure titanium is good, especially in the annealed state, although its crystal structure is α in a compact hexagonal system. This phenomenon is due to the c/a ratio being close to the theoretical value of 1.633, which ensures a greater number of slip planes. The ductility of titanium is largely dependent on the dissolved amount of hydrogen, as well as oxygen and nitrogen. For the highest standardized purity for medical use (class 1, minimum 99.5% Ti), the elongation at break reaches 30%, and the reduction at break exceeds 35% at room temperature. A typical application benefiting from this ductility is surgical staples, used for viscerosynthesis or tissue clips. Regarding chemical properties, titanium is a highly reactive metal (immediately following aluminum in the activity series of metals). It reacts intensely in contact with gases, especially at high temperatures. It burns when heated in the atmosphere at temperatures above 610–650°C and even burns in pure nitrogen above 800°C, being one of the only elements to exhibit such behavior. This requires that the machining of precision parts that need an impurity-free surface, such as medical titanium accessories, is performed only in vacuums more advanced than 10–5 torr or in argon.

The excellent corrosion resistance of titanium is the property that determines its extensive medical use. It withstands all types of water, acid, or salt solutions, exhibiting behavior comparable to platinum in chemical corrosion. It presents some issues in contact with the ClO3 ion and cannot be heated in contact with halogens at temperatures above 550°C due to the so-called salt corrosion.

By combining these excellent properties, titanium proves to be an extremely valuable material for a wide range of applications, offering efficient and durable solutions that surpass the performance of traditional materials.

3. Titanium alloys

Titanium alloys represent a class of advanced materials that enhance and diversify the already remarkable properties of pure titanium, thus expanding the range of applications in which this metal can be used. By alloying titanium with other elements, such as aluminum, vanadium, molybdenum, zirconium, and others, alloys with specific mechanical, thermal, and chemical characteristics can be obtained, tailored to the particular needs of various industries.

Titanium alloys are generally classified into three main categories, based on the crystal structure and alloy composition (**Figure 3**).

- *Alpha (α) alloys*, these alloys contain alpha stabilizers, such as aluminum and oxygen, which improve high-temperature strength and corrosion resistance. They are characterized by good weldability and toughness at low temperatures.

- *Beta (β) alloys* contain beta stabilizers, such as molybdenum, vanadium, and niobium, which lower the beta-alpha transformation temperature, thereby improving ductility and workability. These alloys can be heat-treated to enhance strength.

- *Alpha-Beta (α-β) alloys*, the combination of alpha and beta stabilizers provides a balance between ductility and strength, making these alloys versatile and widely used, especially in the aerospace industry [23, 24].

Alloying elements can significantly influence the properties of titanium, transforming it for specific applications in various industries [25]. Titanium is known for its good properties, but adding alloying elements can improve other characteristics as well. In **Table 2**, some of the most common alloying elements and their effects on titanium are highlighted.

Beta titanium alloys are highly significant in the medical domain owing to their distinctive characteristics:

- the biocompatibility of beta titanium alloys renders them exceptionally suitable for implants and medical devices, as they are well received by the human body and do not elicit unpleasant reactions. It is important for any material that is inserted into the body, as it decreases the likelihood of post-operative problems and rejection of the implant;

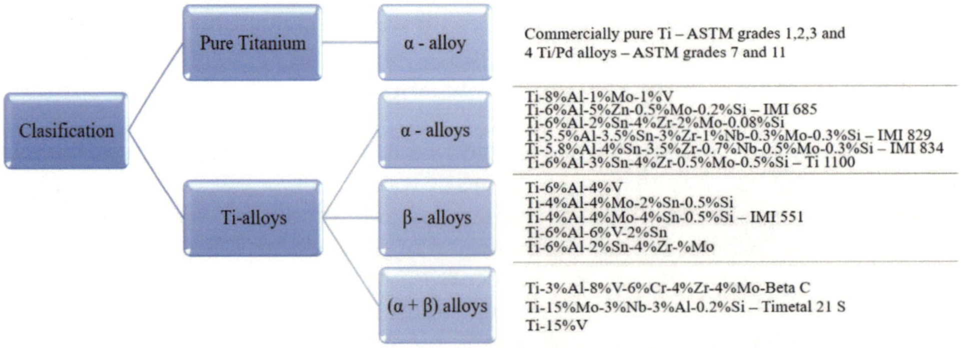

Figure 3.
Titanium alloys classified by metallurgical structure.

Element	Influence on titanium
Aluminum	• the most common alloying element for titanium; • it improves tensile strength and increases the ability to withstand high temperatures; • most titanium alloys contain between 2.5 and 6.5% aluminum.
Vanadium	• usually between 2% and 4%, increases the mechanical strength of titanium without greatly compromising ductility; • titanium alloys containing vanadium are commonly used in the aeronautical industry and in orthopedic implants.
Molybdenum	• added in percentages up to 3%, increases resistance to high temperatures and corrosion; • it also helps maintain strength and ductility at low temperatures.
Zirconium	• is added to improve corrosion resistance, especially in chemically aggressive environments; • it can also be useful in improving the processability of alloys.
Tin	• is used to improve corrosion resistance and increase the strength and hardness of titanium alloys.
Nickel	• improve corrosion resistance in reducing environments and to increase mechanical strength.
Palladium and platinum	• are used in small amounts to improve corrosion resistance in highly corrosive environments such as those in the chemical and petrochemical industries.

Table 2.
Influence of alloying elements on the properties of titanium.

- beta titanium alloys provide exceptional corrosion resistance in biological settings, such as bodily fluids. Implants composed of these alloys exhibit long-term stability and resistance to degradation, guaranteeing both durability and lifespan. Corrosion resistance has an important role for durable implants like hip and knee replacements;

- as they possess excellent mechanical strength and changeable hardness that can be enhanced with heat treatments. As a result, they are capable of enduring significant amounts of pressure and operating effectively when subjected to physical strain. This quality makes them well-suited for use in orthopedic and dental implants, which must endure the forces generated during movement and chewing.

- they are utilized in a diverse range of medical applications: orthopedic implants include prostheses for the hip and knee, as well as plates and screws used for bone stabilization; dental implants consist of root implants and other prosthetic components; surgical instruments refer to equipment that are both lightweight and resistant to corrosion; vascular stents, on the other hand, include coronary stents and other devices used in cardiovascular procedures.

The development and use of titanium alloys continue to evolve, with ongoing research aimed at creating new compositions that offer enhanced performance for specific applications, opening new horizons in engineering and technology.

4. Applications of titanium alloys in the aerospace and maritime industries

In the aerospace industry, materials must meet rigorous criteria for performance and safety. Titanium alloy has been extensively utilized in various industries, including aerospace, military, chemical, medical, and offshore oil, due to its exceptional performance benefits. Often referred to as "space metal" "marine metal" and "smart metal", titanium alloy is a new type of structural metal material pivotal in developing high-tech defense weapons and equipments. Many global powers recognize it as a strategic structural metal for the twenty-first century, essential to their military development. With the rapid and sustainable growth of national economies, the demand for titanium alloys in the aerospace and armament sectors in China has been increasing annually by 20 to 30%.

The United States developed the first titanium alloy (Ti-13 V-11Cr-3Al) in the 1950s that was truly used for flight, particularly in high-speed early warning aircraft. By the 1960s, titanium alloys had become widespread in military aeroengines and large-body jets like the Boeing 747. In the 1970s, titanium alloys accounted for about 80% of the total U.S. titanium alloy market. The 1980s and 1990s saw significant growth in the use of titanium and its alloys in aircraft in Europe and Russia, with Japan also showing a yearly increase in its use in aircraft.

Examples of titanium alloys used in the aerospace industry:

- Ti-6Al-4 V (Grade 5) is the most widely used titanium alloy in the aerospace industry due to its excellent combination of strength, ductility, and corrosion resistance. It is used in manufacturing aircraft structures, engine components, survival systems, and landing gear.

- Ti-6Al-2Sn-4Zr-2Mo, this alloy is valued for its high resistance to elevated temperatures and is used in aircraft engine components such as turbine disks and casings.

- Ti-3Al-2.5 V (Grade 9), this alloy is lighter and has better corrosion resistance compared to Ti-6Al-4 V. It is used in manufacturing tubes for hydraulic systems and other structural applications of aircraft that require good strength and reduced weight.

- Ti-5Al-2Sn-2Zr-4Mo-4Cr (Beta C), this beta alloy is known for its exceptional strength and ability to resist stress cracking. It is used in manufacturing aircraft structural components and in applications requiring high corrosion and cracking resistance.

- Ti-15 V-3Cr-3Al-3Sn (Grade 38) is a beta metastable titanium alloy, valued for its excellent formability at low temperatures and corrosion resistance. It is used in the production of cold-formed parts and in applications where good corrosion resistance and ease of fabrication are required.

In the fabrication of aircraft and spacecraft, particularly in situations where weight is a critical factor, titanium alloys are utilized to production of jet engine parts, including as compressors, turbines, and casings, because of their outstanding ability to withstand high temperatures and resist corrosion. Titanium alloys offer both strength and lightweight, making them advantageous for many aircraft components such as skeletons, fuselage panels, and fasteners. Within the space sector, these components are utilized to endure and tolerate drastic fluctuations in temperature and pressure. They provide exceptional performance and reliability even in the most challenging operational circumstances. Titanium plays an important role in the construction of an aviation engine by providing essential components such as fan blades, fan case, fan disk, and low-pressure compressor blade/stator vane/disk [26, 27].

Titanium and its alloys are extensively used in marine industries due to their distinctive characteristics, such as superior strength, low weight, and extraordinary resistance to corrosion caused by seawater. These materials have a higher level of resistance to corrosion compared to various metals, including aluminum and stainless steels. As a result, they are often selected in harsh settings where they are exposed to salt water and significant temperature fluctuations. Titanium alloys exhibit exceptional mechanical strength even at elevated temperatures, making them highly desirable for use in engine components.

But the use of titanium is valuable in the marine industry not only for their resistance to corrosion, but also for other properties that make them ideal, such as:

- *Corrosion resistance*, one of the most outstanding advantages of titanium alloys is their superior corrosion resistance, especially in environments exposed to chlorides such as seawater. It resists erosion, pitting corrosion and stress cracking, common problems in marine metal construction.

- *Strength-to-weight ratio*, these alloys offer one of the highest strength-to-weight ratios of all metals, making them extremely effective for lightweight structures and components, such as ship superstructures, that require materials that does not add significant extra weight.

- *High-temperature resistance*, titanium alloys maintain their strength and structural integrity at much higher temperatures. This property is useful for marine applications that may be exposed to intense heat sources or temperature variations.

- *Biocompatibility*, titanium is highly resistant to corrosion by marine microorganisms. This feature reduces the need for maintenance and cleaning, extending the life of marine structures.

Among the main applications of titanium in the maritime sector are [28]: *in shipbuilding*, titanium is used in parts of ships exposed to corrosive environments, such as propellers, propeller shafts, and the hull material of certain specialized ships. Its resistance to saltwater corrosion makes it an ideal material for these applications, extending the life of these parts and reducing maintenance costs; *boat hulls and components* are found in the construction of yachts, speedboats and other watercraft, and titanium can be used for structural parts, including the keel and mooring components, to increase strength and reduce maintenance; *desalination plants*, titanium is used in the construction of desalination plants, which turn seawater into drinking water. Its components, especially those involved in heat exchange processes, benefit from titanium's corrosion resistance; *underwater applications*, because of its outstanding corrosion resistance and structural integrity, titanium is often used in underwater equipment and vehicles, including submarines, remotely operated underwater vehicles (ROVs) and underwater instrument housings.

Despite the significant advantages, titanium alloys also have some disadvantages. The high cost is one of their main disadvantages, given the price of the raw material and the complexity of the production process. Machining titanium alloys can also be difficult and expensive, as they require special tools and advanced machining techniques. In addition, titanium can suffer from embrittlement at cryogenic temperatures, limiting its use in space applications that require exposure to extremely cold conditions.

5. Applications of titanium alloys in the medical field

Titanium enjoys remarkable success in medical applications, owing to its exceptional biocompatibility, superior mechanical properties, and high resistance to corrosion. These characteristics make it ideal for various medical devices, including joint implants, bone screws, dental implants, and spinal fixation devices. The adaptability of titanium alloys to the human body minimizes the risk of rejection and increases the durability and success rate of medical implants, thus contributing to improved patient outcomes.

Titanium and its alloys are widely used in the medical field (**Figure 4**), primarily for manufacturing orthopedic implants, such as artificial knee and hip joints, plates, and screws for bone fracture fixation. Titanium is also chosen for dental implants due to its excellent biocompatibility and osseointegration capacity. In cardiology, titanium is used in stents and heart valves. Due to its corrosion resistance and anti-allergic properties, titanium is ideal for long-term applications within the human body [29].

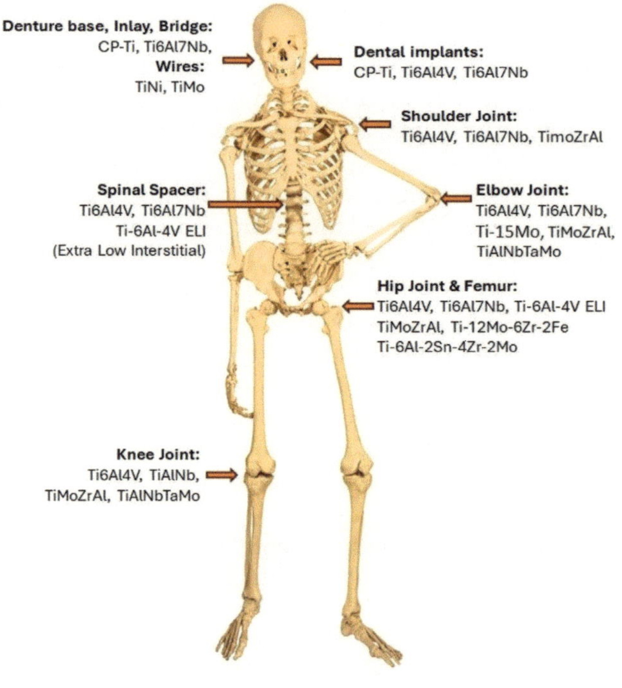

Figure 4.
Titanium alloys used in medical devices throughout the entire human body.

Titanium began to be used in medical applications in the 1950s, owing to its exceptional properties of biocompatibility and corrosion resistance. Initially, it was used in dental prostheses, and its use gradually expanded to other types of implants, such as orthopedic and cardiac implants. Due to its ability to integrate well with human bone (osseointegration), titanium has become the preferred material for many types of medical implants, significantly improving the quality of life for patients.

Titanium exhibits exceptional biological compatibility, contributing to the process of osseointegration, where bone tissue adheres directly to the implant surface without causing chronic inflammation. This is due to the formation of a layer of titanium oxide on the metal surface, which not only provides corrosion resistance but also promotes bone adhesion [1].

For medical applications, some of the most common titanium alloys are as follows:

- Ti-6Al-4 V (Titanium Grade 5), probably the most widely used titanium alloy in the medical field. It contains 6% aluminum and 4% vanadium and offers good mechanical strength and corrosion resistance. It is used for orthopedic implants such as hip or knee prostheses.

- Pure titanium (Titanium Grade 1 and Grade 2) is used especially for surgical implants, such as plates or screws used in reconstructive or maxillofacial surgery. It is preferred because it is biocompatible and corrosion-resistant.

- Ti-6Al-7Nb (Titanium Grade 7), this alloy contains niobium and is used especially for implants that require superior corrosion resistance. It is also biocompatible and has elasticity similar to bones, making it suitable for certain surgical applications.

- A fascinating aspect of titanium is its alloys with nickel, such as Nitinol-55, which exhibits shape memory; these can regain their original shape when heated above a certain temperature. This property is leveraged in various medical applications, including dentistry, cardiac surgery, and orthopedics, due to their excellent ductility at low temperatures, biocompatibility, and corrosion resistance [30, 31].

Biologically, titanium is non-magnetic, does not interfere with magnetic fields, and promotes regenerative processes, attracting calcium ions and favoring the formation of hydroxyapatite around the implant. Osseointegration represents another major advantage, with titanium establishing a strong bond with the surrounding bone.

However, titanium also has disadvantages, including a relatively low shear and wear resistance, as well as difficulties in the manufacturing process. Despite these challenges, the positive qualities of titanium, including its corrosion resistance, biocompatibility, and osseointegration capacity, make it the preferred material for numerous medical applications, from dental and orthopedic implants to artificial hearts and other surgical devices.

It is interesting to note that although titanium is widely used in medicine, the amount of titanium released into the body from an implant is thousands of times less than what is naturally metabolized by the human body daily, underscoring its safety and biological irrelevance in the context of implants [32, 33].

Therefore, the exceptional qualities of titanium, both in terms of its physical–mechanical and biological aspects, make it an essential material in the medical field, significantly contributing to the success of surgical interventions and improving the quality of life for patients.

New alloys from the TiMoZrTaSi system, such as Ti20Mo7Zr15TaxSi (where x = 0.5, 0.75, 1%) [4, 18], show significant potential for medical applications. These alloys are carefully engineered to possess essential properties for medical devices, including biocompatibility, low toxicity, and mechanical attributes similar to those of bone. Through experimental production, these alloys have demonstrated changes in mechanical properties with the addition of silicon, resulting in reduced hardness and a slight increase in the modulus of elasticity as silicon content increases.

Cytocompatibility assessments conducted on fibroblasts and osteoblasts revealed no adverse effects on cell proliferation or morphology after incubation with these alloys. In vivo studies confirmed their excellent biocompatibility and ability to facilitate bone remodeling without impeding new bone formation.

Similarly, alloys from the Ti15Mo7Zr15TaxSi system were developed, with a focus on their chemical and mechanical properties, as well as biocompatibility. These alloys showed high production efficiency and maintained the desired chemical composition. The addition of silicon improved mechanical properties, leading to increased hardness and modulus of elasticity.

In vitro cytocompatibility tests showed no cytotoxic effects, with a slight increase in cell viability at higher silicon concentrations. Further in vivo studies demonstrated enhanced osseointegration after implantation, along with significant bone remodeling activity in the peri-implant areas.

Additionally, biocompatible Ti-based alloys from the Ti20MoxSi system were developed by adjusting the silicon content while keeping molybdenum levels constant. These alloys exhibited improved mechanical properties with the incorporation of silicon, as evidenced by a decrease in modulus of elasticity and hardness.

Cytocompatibility tests confirmed positive interactions with both fibroblasts and osteosarcoma cells, without affecting cell growth or morphology. In vivo evaluations supported their compatibility with surrounding tissue and successful osseointegration, accompanied by notable bone remodeling near implant sites. This research highlights the potential of newly engineered titanium alloys for medical applications, demonstrating excellent biocompatibility, mechanical properties, and the ability to support bone integration and remodeling.

Furthermore, Ti-based alloys are attractive for biomedical applications due to their superior mechanical properties, corrosion resistance, and biocompatibility. Another study provides a comparative analysis of four novel Ti-based alloys designed for various biomedical implant applications: Ti15Mo7Zr5Ta, Ti15Mo7Zr15Ta, Ti15Mo0.5Si, and Ti15Mo1Si, alloys that incorporate non-toxic elements. The research included microstructural examinations, indentation tests, Vickers hardness assessments, X-ray diffraction (XRD) analysis, and evaluations of corrosion resistance. These analyses revealed attributes superior to those of many commercial implant materials, with Young's modulus closely matching that of human bone.

There is a continuous global interest in advancing alloy research for medical and biomedical applications, aiming to improve traditional implant manufacturing technologies and biomaterial synthesis. This pursuit strives to introduce a new generation of multifunctional implants with long-lasting performance.

Biomaterials are intended to mimic or closely interact with living tissues, necessitating properties similar to human bone. The Young's modulus of Ti-based alloys, ranging from 19.82 to 69.02 GPa, suits them well for implantology applications.

Surface analyses have revealed titanium oxidation, which forms an adherent oxide layer that passivates the alloy, leading to relatively low corrosion rates across alloys with uniform compositions. The inclusion of tantalum has been found to reduce both the modulus of elasticity and the corrosion rate, while silicon has shown beneficial effects at low percentages (below 0.5%) [34].

In conclusion, the characteristics of titanium alloys depend on the alloying elements, influencing their mechanical properties and biocompatibility. Elements such as molybdenum, zirconium, tantalum, and silicon, when combined with titanium, confer advantageous properties, promising a wide range of medical applications, including for dental and orthopedic implants.

Advancements in titanium alloys have been marked by innovative treatments and coatings, aiming to optimize their application in the medical field, particularly for implants. Among these, the application of zirconia coatings and hydroxyapatite/tricalcium phosphate via biomimetic methods have demonstrated promising enhancements in biocompatibility and mechanical properties [35].

Zirconia coatings on substrates such as Ti15Mo, Ti15Mo0.5Si, Ti15Mo0.75Si, and Ti15Mo1.0Si showcased uniform morphologies without microcracks, highlighting the presence of β-Ti and ZrO2 phases with a tetragonal crystalline structure. These coatings, notably devoid of un-melted zirconia compounds, exhibited an improved modulus of elasticity, significantly enhancing the alloys' suitability for orthopedic implants by mimicking the mechanical properties of human bone more closely.

Furthermore, the impact of heat treatments on Ti-Mo-Zr-Ta alloys revealed important insights into microstructural optimization and mechanical property enhancement. Focusing on superficial hardening, these treatments have led to better wear behavior, fatigue, and corrosion resistance. The formation of β-type structures, aided by the precise inclusion of β stabilizing elements like molybdenum, tantalum, and silicon, results in a refined and evenly distributed microstructure, improving both hardness and elasticity modulus [36].

Additionally, the coating of HA/tricalcium phosphate on titanium surfaces using the biomimetic method has emerged as a groundbreaking advancement. This technique facilitates the formation of a bone-like apatite layer in simulated physiological conditions, enhancing the bioactivity and bacteriostatic properties of the implants. These coatings, characterized by nanometric crystals similar to natural bone, promise a leap forward in the biocompatibility and integration of titanium-based implants [37].

Surface wettability analysis further underscores the biocompatible potential of these treated alloys, with specific compositions exhibiting hydrophilic characteristics conducive to better interaction with living tissues.

These advancements underscore a multidirectional approach to enhancing titanium alloys for medical applications. By focusing on coatings that improve biocompatibility and mechanical properties, alongside heat treatments that optimize microstructure and wear resistance, the field is moving toward the development of more durable, reliable, and biologically harmonious implants.

Current trends in the development of alloys for medical applications focus on enhancing biocompatibility and reducing adverse effects in the body. Alloys with elements such as zirconium and tantalum, known for their reduced corrosion and minimal reactions with tissues, are being explored. Additionally, shape memory alloys are being developed for specific applications that require adaptability to the conditions of the human body. Another area of interest is the creation of nano-modified alloy surfaces to improve osseointegration and reduce the risk of infections.

6. Applications of titanium alloys in the automotive industry

Titanium, owing to its distinctive characteristics, has been utilized in significant capacities not only in the medical domain but also in the automotive sector. The utilization of this material in these industries is grounded on its remarkable resistance to corrosion, exceptional ratio of strength to weight, and resilience to high temperatures.

Titanium is highly prized in the automotive sector for its capacity to decrease the weight of vehicles, hence enhancing fuel efficiency and dynamic performance. Despite the expensive nature of titanium, it is frequently employed in the parts of high-performance automobiles, supercars and in the realm of auto racing, where its benefits outweigh the expense. Some specific applications of this material include manufacturing important engine components such as valves, valve springs, valve spring retainer seats, and connecting rods. The lower weight and high strength of titanium contribute to enhanced engine performance and efficiency, benefiting these components.

Chassis components are utilized in various applications such as springs, exhaust systems, axle shafts, and fasteners. These applications utilize titanium's corrosion resistance, fatigue resistance, and strength.

Other applications include suspension springs, piston bolts, turbocharger rotors, fasteners, wheel nuts, bumper brackets, door beams, brake caliper pistons, bolts, clutch discs, pressure plates, and gear shift knobs. These components have the potential to greatly decrease the total weight of the vehicle, hence enhancing its performance and fuel efficiency.

Two major obstacles in utilizing titanium in the automobile sector are its exorbitant price and the complexities associated with the material and manufacturing processes. The primary obstacle impeding the extensive utilization of titanium in the automobile industry has been its high price. The process of titanium processing entails the melting, manufacturing, and utilization of various materials. The exorbitant expense of alloying elements also adds to the overall costs. The extraction and processing of titanium are challenging due to its high melting point and its reactivity with elements such as oxygen, hydrogen, nitrogen, and carbon. The conventional Kroll process utilized for manufacturing sponge titanium is characterized by high energy consumption, a lengthy production cycle, and the utilization of costly magnesium as a reducing agent.

In order to decrease manufacturing expenses, contemporary methods such as the utilization of titanium scrap and swarf for repetitive production have been implemented. These methods have been identified as efficient means to decrease the expenses associated with acquiring raw materials. Each incremental 1% rise in the utilization of titanium scrap results in a corresponding reduction of 0.8% in the manufacturing expenses associated with ingots.

7. Challenges and limitations

Although titanium is a material with many advantages, there are also challenges and limitations in its use, both in the previously mentioned industries and in general applications. These limitations are often related to its cost, processing difficulties, and performance considerations under certain conditions (**Figure 5**).

Despite these challenges and limitations, continuous innovations in technology and manufacturing processes contribute to cost reduction and improved efficiency in titanium processing. Additionally, the development of new alloys and surface treatments expands the scope of titanium applications, overcoming some of its initial limitations and paving the way for innovative and high-performance uses in various industries.

8. Future trends and innovations

Advancements in technology and ongoing research in materials science have paved the way for new trends and innovations in titanium usage. These developments

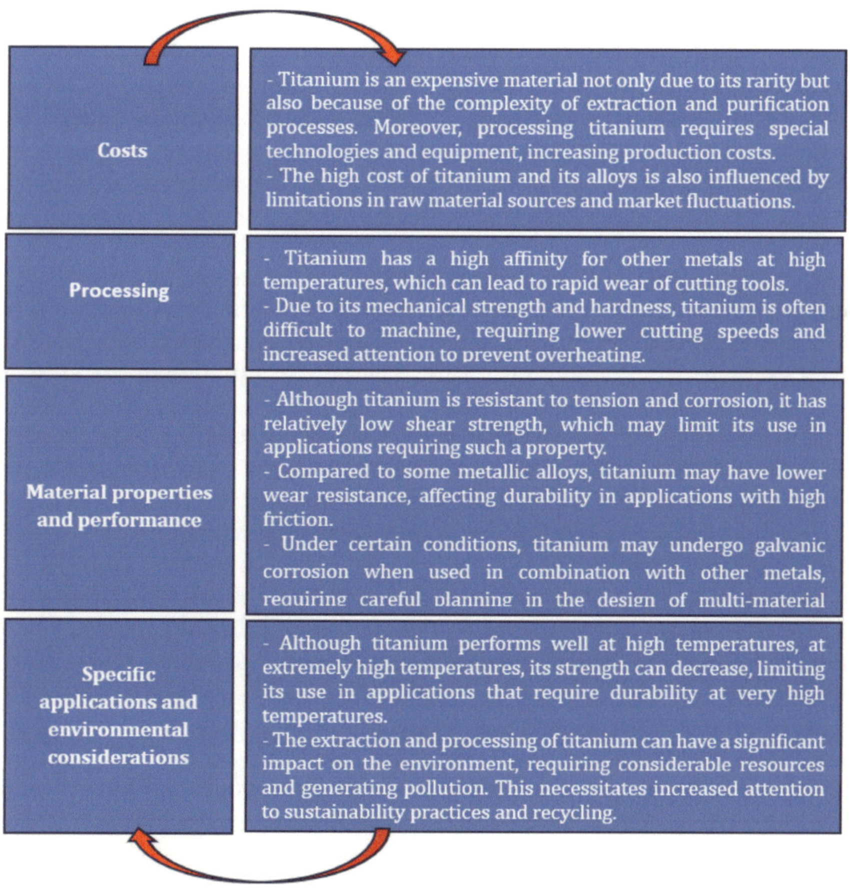

Figure 5.
Limitations of Ti alloys.

promise to overcome some of the current limitations and expand the applicability of titanium across various industries.

Researchers are working on the development of new titanium alloys that offer an improved balance of strength, ductility, and corrosion resistance at varying temperatures. These new alloys are designed to outperform in specific applications, such as highly corrosive environments or high-temperature settings, thereby extending the use of titanium into new domains like advanced aerospace industry and renewable energy applications.

Innovations in materials processing, such as additive manufacturing (3D printing) and superplastic forming, are transforming how titanium can be utilized. 3D printing enables the creation of complex titanium components with minimal material waste and at reduced costs compared to traditional methods. This opens up new possibilities in the design of customized products, ranging from medical implants to specific parts for vehicles and aircraft.

Characteristics of 3D printed titanium alloys can exhibit exceptional mechanical strength, characterized by high tensile strength and favorable fatigue resistance. However, the specific properties may be subject to variation depending on the printing conditions and post-processing procedures employed. The microstructure of titanium alloys produced using 3D printing can exhibit greater complexity compared to alloys manufactured using traditional methods. The high-speed cooling rates and step-by-step manufacturing can lead to intricate microstructures and distinctive grain formations. A notable benefit of 3D printing is its capacity to create intricate geometries that are challenging or unattainable using traditional techniques. This encompasses lattice structures as well as interior channels. Generally, post-processing is necessary for 3D printed components to attain the intended surface finish and dimensional tolerances. This can encompass processes such as machining, polishing, or other forms of treatment [35–39].

There is a strong push toward improving titanium extraction and processing processes to make them more environmentally friendly. Titanium recycling is becoming a priority, with the development of technologies that enable efficient recovery and reuse of titanium waste. These innovations reduce environmental impact and make titanium more accessible and sustainable.

Titanium holds significant potential in the field of renewable energy, particularly in wind turbines and marine applications, due to its corrosion resistance in saline environments and material fatigue resistance. Developing titanium components for these systems can enhance their efficiency and durability, contributing to a more sustainable energy transition.

Innovations in the field of biomedicine are exploring the use of titanium in smart implants, which can monitor healing and release drugs at the implant site. These technologies promise to improve treatment outcomes and provide personalized solutions to patients.

The development of composite and hybrid materials that integrate titanium with other materials, such as carbon fibers or advanced ceramics, offers new perspectives for creating ultra-lightweight and high-performance components. These combine the superior properties of titanium with those of other materials to achieve improved performance in specific applications.

These trends and innovations underscore the titanium's potential to play an even more significant role in future technology. By overcoming current challenges and exploiting new possibilities, titanium remains at the forefront of advanced materials development, offering innovative solutions for a wide range of industrial and technological applications.

Ongoing research in titanium alloys aims to overcome existing limitations by developing new compositions and processing techniques to reduce costs and improve material properties. Innovations in additive manufacturing (3D printing) promise to revolutionize how titanium components are produced, offering the ability to create complex shapes with material efficiency and reduced costs. Additionally, research in surface treatments and intermetallic alloys opens up new possibilities for the use of titanium in even more demanding conditions.

9. Conclusions

Titanium, owing to its unique combination of remarkable physical and chemical properties, has revolutionized many fields, from medicine and aerospace industry to automotive and maritime applications. However, its use comes with specific challenges, including high production costs, processing difficulties, and limitations in certain applications due to material properties. Despite these obstacles, continuous innovations in alloys, processing techniques, and recycling promise to further expand the uses of titanium, surpassing current limitations and opening new horizons for this versatile material.

Advancements in additive manufacturing, the development of new alloys, and advanced processing techniques, as well as the focus on sustainability and recycling, have the potential to reduce costs and make titanium more accessible for a wider range of applications. Additionally, exploring titanium's potential in emerging fields such as renewable energy and biomedicine highlights its role in addressing some of the most pressing challenges of modern society.

In conclusion, titanium continues to be a frontier material, with significant potential to contribute to innovations in design, engineering, and technology. As research progresses and technologies develop, titanium is expected to maintain its position as an essential material, opening up new possibilities for the future of technological and industrial development.

Acknowledgements

"This work was supported by a grant of from the Ministry of Research, Innovation and Digitization, CNCS/CCCDI—UEFISCDI, project number ERA-NET-ERAMIN-3-Cool&SmartTit-1, contract no. 8/2024 within PNCDI IV."

Conflict of interest

The authors declare no conflict of interest.

Author details

Madalina Simona Baltatu[1], Petrică Vizureanu[1,2*], Andrei Victor Sandu[1,3,4,5], Dragos Cristian Achitei[1], Manuela Cristina Perju[1], Dumitru Doru Burduhos-Nergis[1] and Marcelin Benchea[6]

1 Faculty of Material Science and Engineering, "Gheorghe Asachi" Technical University of Iasi, Iasi, Romania

2 Technical Sciences Academy of Romania, Bucharest, Romania

3 Romanian Inventors Forum, Iasi, Romania

4 Academy of Romanian Scientists, Splaiul Independentei, Bucharest, Romania

5 National Institute for Research and Development in Environmental Protection, Bucharest, Romania

6 Faculty of Mechanical Engineering, "Gheorghe Asachi" Technical University of Iasi, Iasi, Romania

*Address all correspondence to: peviz2002@yahoo.com

References

[1] Ratner BD, Bryant SJ. Biomaterials: Where we have been and where we are going. Annual Review of Biomedical Engineering. 2004;**6**:41-75

[2] Geetha M, Singh AK, Asokamani R, Gogia AK. Ti based biomaterials, the ultimate choice for orthopaedic implants – A review. Progress in Materials Science. 2009;**54**:397-425

[3] Istrate B, Munteanu C, Geanta V, Baltatu S, Focsaneanu S, Earar K. Microstructural analysis of biodegradable Mg-0.9Ca-1.2Zr alloy. IOP Conference Series: Materials Science and Engineering. 2016;**147**:012033

[4] Baltatu MS, Spataru MC, Verestiuc L, Balan V, Solcan C, Sandu AV, et al. Design, synthesis, and preliminary evaluation for Ti-Mo-Zr-Ta-Si alloys for potential implant applications. Materials. 2021;**14**(22):6806

[5] Hildebrand HF. Biomaterials–A history of 7000 years. BioNanoMaterials. 2013;**14**:119-133

[6] Huang J, Best SM. Ceramic biomaterials. In: Tissue Engineering Using Ceramics and Polymers. Netherlands: Woodhead Publishing; 2007. pp. 3-31

[7] Antoniac I, Miculescu M, Mănescu (Păltânea) V, Stere A, Quan PH, Păltânea G, et al. Magnesium-based alloys used in orthopedic surgery. Materials. 2022;**15**:1148

[8] Shahani S. Advanced drug delivery systems: New developments, new technologies Report No. PHM006F. Business Communications Company; 2006

[9] Baltatu MS, Vizureanu P, Goanță V, Tugui CA, Voiculescu I. Mechanical tests for Ti-based alloys as new medical materials. IOP Conference Series: Materials Science and Engineering. 2019;**572**:012029

[10] Bronzino JD, Peterson DR, editors. The Biomedical Engineering Handbook. 4th ed. Boca Raton, Florida: CRC Press; 2015

[11] Verma RP. Titanium-based biomaterial for bone implants: A mini review. Materials Today Proceedings. 2020;**26**:3148-3151

[12] Baltatu MS, Sandu AV, Nabialek M, Vizureanu P, Ciobanu G. Biomimetic deposition of hydroxyapatite layer on titanium alloys. Micromachines. 2021;**12**:1447

[13] Savin A, Vizureanu P, Prevorovsky Z, Chlada M, Krofta J, Baltatu MS, et al. Noninvasive evaluation of special alloys for prostheses using complementary methods. IOP Conference Series: Materials Science and Engineering. 2018;**374**:012030

[14] Baltatu I, Sandu AV, Vlad MD, Spataru MC, Vizureanu P, Baltatu MS. Mechanical characterization and In vitro assay of biocompatible titanium alloys. Micromachines. 2022;**13**(3):430

[15] Istrate B, Munteanu C, Bălțatu M-S, Cimpoeșu R, Ioanid N. Microstructural and electrochemical influence of Zn in MgCaZn biodegradable alloys. Materials. 2023;**16**:2487. DOI: 10.3390/ma16062487

[16] Bita AI, Antoniac A, Cotrut C, Vasile E, Ciuca I, Niculescu M, et al. In vitro degradation and corrosion evaluation of Mg-Ca alloys for biomedical applications. Journal of Optoelectronics and Advanced Materials. 2016;**18**(3-4):394-398

[17] Kokubo T, Yamaguchi S. Bioactive metals prepared by surface modification: Preparation and properties. In: Eliaz N, editor. Applications of Electrochemistry in Biology and Medicine I. Midtown Manhattan, New York City: Springer Science+Business Media; 2011. pp. 377-421

[18] Spataru M-C, Cojocaru FD, Sandu AV, Solcan C, Duceac IA, Baltatu MS, et al. Assessment of the effects of Si addition to a new TiMoZrTa system. Materials. 2021;**14**:7610

[19] Sidambe AT. Biocompatibility of advanced manufactured titanium implants—A review. Materials. 2014;7:8168-8188

[20] Prasad K, Bazaka O, Chua M, Rochford M, Fedrick L, Spoor J, et al. Make better, safer biomaterials. Nature. 2016;**540**:335

[21] Niinomi M. Mechanical properties of biomedical titanium alloys. Materials Science and Engineering A. 1998;**243**:231-236

[22] Wang M, Guo L, Sun H. Manufacture of biomaterials. In: Narayan R, editor. Encyclopedia of Biomedical Engineering. The Netherlands: Elsevier; 2019. pp. 116-134

[23] Chen Q, Thouas GA. Metallic implant biomaterials. Materials Science & Engineering R: Reports. 2015;**87**:1-57

[24] Salihu Sani A et al. Classification, properties and applications of titanium and its alloys used in automotive industry- A review. American Journal of Engineering Research (AJER). 2019;**8**(8):92-98

[25] Boyer RR. Titanium for aerospace: Rationale and applications. Advanced Performance Materials. 1995;**2**:349-368. DOI: 10.1007/BF00705316

[26] Inagaki I, Takechi T, Ariyasu YSN. Application and Features of Titanium for the Aerospace Industry. Jinan City: Nippon Steel & Sumitomo Metal Technical Report No. 106. 2014. pp. 22-27

[27] Inagaki I, Takechi T, Shirai Y, Ariyasu N. Application and Features of Titanium for the Aerospace Industry, Nippon Steel & Sumitomo Metal Technical Report No. 106. 2014

[28] Gomez-Gallegos AA, Mandal P, González D, Zuelli N, Blackwell P. Studies on titanium alloys for aerospace application. Defect and Diffusion Forum. 2018;**385**:419-423

[29] Sarraf M, Rezvani Ghomi E, Alipour S, et al. A state-of-the-art review of the fabrication and characteristics of titanium and its alloys for biomedical applications. Bio-Design and Manufacturing. 2022;**5**:371-395. DOI: 10.1007/s42242-021-00170-3

[30] Prasad K, Bazaka O, Chua M, Rochford M, Fedrick L, Spoor J, et al. Metallic biomaterials: Current challenges and opportunities. Materials. 2017;**10**(8):884

[31] Virtanen S. Corrosion of biomedical implant materials. Corrosion Reviews. 2008;**26**:147-171

[32] Liu Z, Liu X, Ramakrishna S. Surface engineering of biomaterials in orthopedic and dental implants: Strategies to improve osteointegration, bacteriostatic and bactericidal activities. Biotechnology Journal. 2021;**16**:2000116

[33] Todros S, Todesco M, Bagno A. Biomaterials and their biomedical applications: From replacement to regeneration. PRO. 1949;**2021**:9. DOI: 10.3390/pr9111949

[34] Attarilar S, Salehi MT, Fadhalah K. Functionally graded titanium implants:

Characteristic enhancement induced by combined severe plastic deformation. PLoS One. 2019;**14**:e0221491

[35] Zhang X, Liu S, Liu Y, Guo H, Shi W. Titanium alloy fabricated by additive manufacturing for medical applications: Obtaining, characterization and application—Review. Metals. 2023;**13**:462. DOI: 10.3390/met13030462

[36] Schutz R, Watkins H. Recent developments in titanium alloy application in the energy industry. Materials Science and Engineering A. 1998;**243**:305-315

[37] Xiong J, Yang J, Zhang J, Guo J. Research progress of metal additive manufacturing

[38] Furuta T. Chapter 4 – Automobile applications of titanium. In: Froes F, Qian M, Niinomi M, editors. Titanium for Consumer Applications. The Netherlands: Elsevier; 2019. pp. 77-90. ISBN 9780128158203. DOI: 10.1016/B978-0-12-815820-3.00006-X

[39] Parthasarathy L, Starly B, Raman S, Christensen A. Mechanical evaluation of porous titanium (Ti6Al4V) structures with electron beam melting (EBM). Journal of the Mechanical Behavior of Biomedical Materials. 2010;**3**:249-259

Chapter 2

Mechanical Properties and Performance of Titanium-Based Alloys Used in Aerospace Applications

Amal Mohammed Abdulrahman,
Abdul Mohshen Sharif Ullah Siddique and Rayanah Barnawi

Abstract

This chapter in this book will focus on the mechanical properties, including strength, toughness, and fatigue resistance, of titanium-based alloys and their significance in aerospace applications. It will discuss several types of titanium alloys and explore the unique characteristics of these alloys, such as high strength-to-weight ratio, corrosion resistance, and excellent high-temperature performance. The chapter also will discuss specific challenges and considerations in designing and manufacturing components using titanium-based alloys for aerospace applications, highlighting the benefits and limitations of these materials. Additionally, it will provide case studies and examples of successful applications in the aerospace industry, showcasing the uniqueness and effectiveness of titanium-based alloys in this field.

Keywords: advanced materials, polymer composites, titanium applications, titanium alloys, aerospace

1. Introduction

1.1 Historical background and development of titanium alloys for aerospace

The story of titanium alloys in aerospace begins with the discovery of titanium as a chemical element in the late eighteenth century [1]. Titanium alloys for aerospace span several decades, marked by significant advancements in material science, manufacturing processes, and the pursuit of lightweight and high-performance materials [2, 3]. In this detailed account, we will explore key milestones, challenges, and achievements that have shaped the use of titanium alloys in aerospace applications.

During World War II, the aerospace industry recognized the potential of titanium alloys due to their high strength-to-weight ratio and corrosion resistance [4]. However, it was not until the early twentieth century that researchers start exploring its potential applications. The processing and manufacturing techniques have played a crucial role in the widespread adoption of titanium alloys in aerospace. These techniques are hot and cold forming, forging, precision machining, and advanced joining methods such as electron beam welding, friction stir welding, and Kroll process [5]. Kroll process developed in 1930s by William Kroll, a commercially viable method for extracting titanium from its ore [6, 7]. This breakthrough laid the foundation for the future development of titanium alloys and has allowed for the fabrication of complex titanium components with enhanced precision, reliability, and reduced production costs [5].

However, widespread use was limited by the high cost and challenges associated with the extraction and processing of titanium [8]. After the war, efforts to overcome these obstacles led to the establishment of commercial-scale production facilities and advancements in manufacturing techniques.

Throughout the 1950s and 1960s, researchers focused on improving mechanical properties of titanium alloys through alloying and heat treatment methods [9]. According to previous studies, there are many types of titanium alloys, such as titanium aluminum tin zirconium molybdenum alloy (Ti-6Al-2Sn-4Zr-2Mo), titanium aluminum tin alloy (Ti-5Al-2.5Sn), and titanium aluminum vanadium alloy (Ti-3Al-2.5V). One significant breakthrough came with the development of titanium aluminum vanadium alloy (Ti-6Al4V), which became the most widely used titanium alloy in aerospace applications [10]. The combination of this alloy offered an excellent strength, corrosion resistance, and weldability [11].

In the 1960s and 1970s, the aerospace industry recognized the unique advantages of titanium alloys, such as their lightweight nature, high strength, and resistance to corrosion and fatigue. Aircraft manufacturing began incorporating titanium components in air frames, engine systems, landing gears, and other critical structures [12]. Notable examples include the Boeing 747 and Concorde, which utilized titanium extensively [13]. The percentage of titanium in different types of aircraft is shown in **Figure 1**.

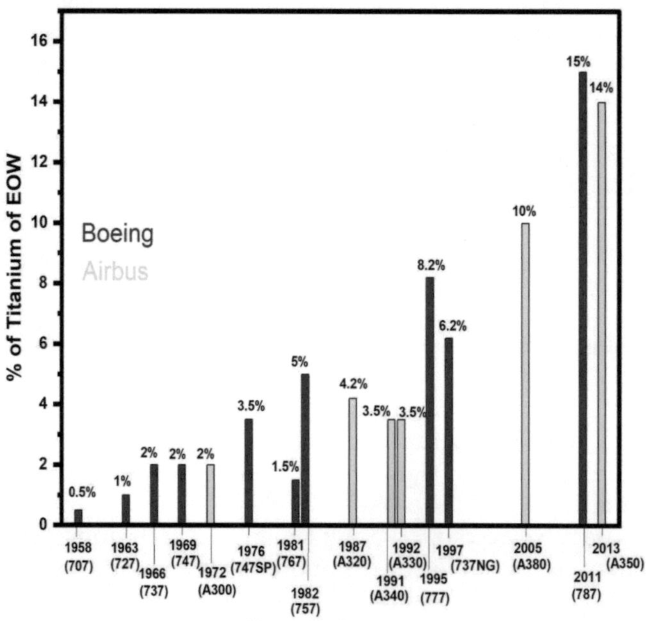

Figure 1.
Share of Ti in Boeing and Airbus aircraft as a percentage of operating empty weight (OEW) (edited and quoted from Ref. [14]).

In case of research and innovation, the development of titanium alloys for aerospace continues to evolve with ongoing research and innovation. Efforts are focused on improving the properties of existing alloys, developing new alloys with enhanced characteristics, and exploring advanced manufacturing technologies. Additive manufacturing is a promising technology in titanium alloy-based materials with the advantages of creating lightweight materials and reducing greenhouse gas emissions which enhance what is known as operation empty weight (OEW), thus increasing the usage of titanium in Boing and Airbus aircrafts as shown in **Figure 1** [14]. Research is also investigating the use of additive manufacturing (3D printing) to create complex titanium components with reduced material waste and increased design freedom [15].

Looking ahead, the demand for titanium alloys in the aerospace to grow as aircraft manufactures seeks to improve fuel efficiency, reduce emissions, and enhance overall performance. However, challenges remain, including the high cost of raw materials [4], the complexity of processing titanium, and the need for further research to optimize the properties of titanium alloys for specific aerospace application.

2. Types of titanium alloys based on equilibrium at room temperature

The term of alpha (α) and beta (β) titanium refer to different crystal structures that titanium can exhibit at different temperatures. There are five types of titanium based on equilibrium at room temperature called α alloys, near α alloys, β alloys, α + β alloys, and titanium-intermetallic compound.

In the α phase, titanium has a hexagonal close-packed (HCP) crystal structure as shown in **Figure 2**. This phase is stable at temperature below 882°C, thus α titanium alloy has excellent properties which make them commonly used in application of aerospace components, biomedical implants, and chemical processing equipment. β titanium alloy has a body-centered cubic crystal structure (BCC) (**Figure 3**) and stable at temperature above 882°C. For α-β titanium alloy, the alpha phase may have some amount of β phase and likewise for beta phase. However, the β phase exhibit

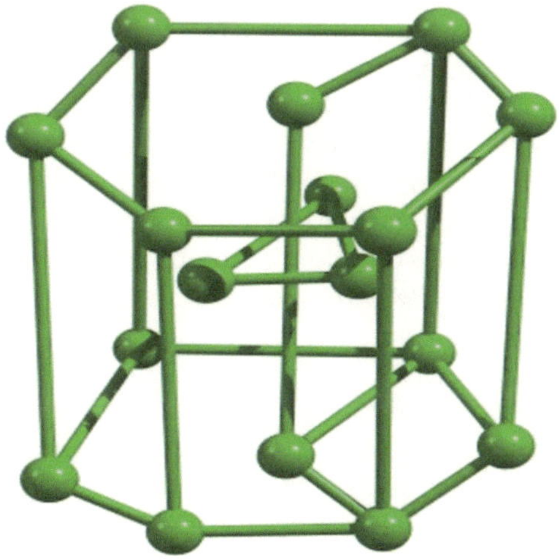

Figure 2.
Titanium hexagonal close-packed crystal structure Ti-(HCP).

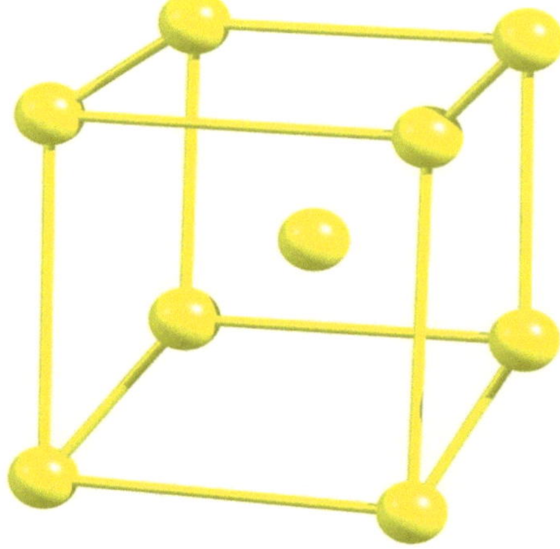

Figure 3.
Titanium body-centered cubic crystal structure Ti-(BCC).

higher strength potential through aging, leading to increased mechanical properties, better deformability, higher diffusivity for interstitial elements, isotopic properties, high density, and lower creek resistance compared to α phase. But generally, β phase of titanium is more expensive than α phase [16, 17].

3. Overview of the significance and demand for titanium-based alloys in aerospace

Characteristics of titanium-based alloys play an important role in aerospace applications duo to their extraordinary properties and performance. Titanium-based alloys exhibit low density and exceptional thermal and mechanical resistance [18]. These properties are result of the unique composition and crystal structure of titanium alloys [19], which can be tailored through alloying and heat treatment processes, making them ideal for various aerospace applications [20]. The demand for titanium-based alloys in aerospace continues to grow due to the following reasons: For lightweight construction, titanium alloys have significantly lower density compared to steel and other metals, such as iron and nickel, resulting in lighter aircraft components [21]. Moreover, the weight reduction contributes to fuel efficiency, increased payload capacity, and improved overall performance [22]. **Table 1** shows the different properties of titanium compared with other metals.

Titanium-based alloys possess excellent strength-to-weight ratios, offering exceptional structural integrity while keeping weight to a minimum. This characteristic is crucial for critical aerospace components that require both strength and lightness, such as airframe structures, engine components, and landing gear [24].

Aerospace applications often involve exposure to harsh environments, including high altitudes, temperature variations and corrosive substances. Titanium-based alloys exhibit exceptional corrosion resistance, ensuring the longevity and reliability of aircraft components, even in challenging operating conditions like high temperatures. For this reason, titanium-based alloys are suitable for components exposed to extreme heat, such as turbine blades and exhaust systems. Their ability to retain strength and integrity at elevated temperatures contributes to enhanced safety and durability [25].

The properties	Ti	Al	Fe	Ni
Melting point [°C]	1670	660	1538	1455
Density [g/cm^3]	4.5	2.7	7.9	8.9
Fatigue resistance	Very high	Medium	Low	High
Elastic modulus [GPa]	115	72	215	200
Reactivity with oxygen	Very high	High	Low	Low
Electrochemical Corrosion resistance	Very high	High	Low	Medium
Metal price	Very high	Medium	Low	High
Tensile strength [MPa]	240	90	50	59
Thermal conductivity [W/mK]	15-22	221-247	68-80	72-92

Table 1.
Comparison of titanium metal properties with different metals [18, 23].

The fatigue resistance of titanium-based alloys is crucial as aircraft components undergo cyclic loading during operation. These alloys demonstrate excellent fatigue resistance [23], reducing the risk of structural failures and ensuring the safety of the aircraft. Meanwhile, properties of titanium-based alloys composites exhibit good compatibility with composite materials, which are increasingly used in modern aerospace design. Their ability to bond effectively with composites allows for the fabrication of hybrid structures [26], combining the advantages of both materials. In addition to these exceptional properties, titanium-based alloys are in high demand in the aerospace industry. As aircraft manufacturers strive for improved performance, fuel efficiency, and safety, the utilization of titanium-based alloys continues to increase across various aircraft systems and components, including engines and fasteners. The growth in the aerospace sector, coupled with ongoing advancements in materials science and manufacturing processes, further drives the significance and demand for titanium-based alloys in aerospace applications.

4. Titanium extraction process

Titanium extraction poses challenges due to its high melting point, chemical reactivity, and the stability of its oxide, titanium dioxide (TiO_2). Early attempts to reduce TiO_2 resulted in impurities that compromised the mechanical properties. Various methods were explored, including electrolysis and reduction with sodium, magnesium, and other substances. The Kroll process, developed in 1937, involving the reduction of titanium tetra chloride ($TiCl_4$) with magnesium, became a widely adopted method for producing high-purity titanium as shown in **Figure 4**. Another method, fused salt electrolysis of $TiCl_4$, showed potential but has not been commercially implemented. The Hunter process, using sodium reduction, confronts challenges such as handling hazards, temperature control, and reductant regeneration. The Kroll process and sodium reduction were used for industrial production of titanium sponge, but some companies have ceased operations. Electrowinning of titanium through fused salt electrolysis of $TiCl_4$ has been studied but not successfully implemented on a commercial scale. The challenges of this method include maintaining effective separation of anolyte and catholyte and ensuring quality in harvesting

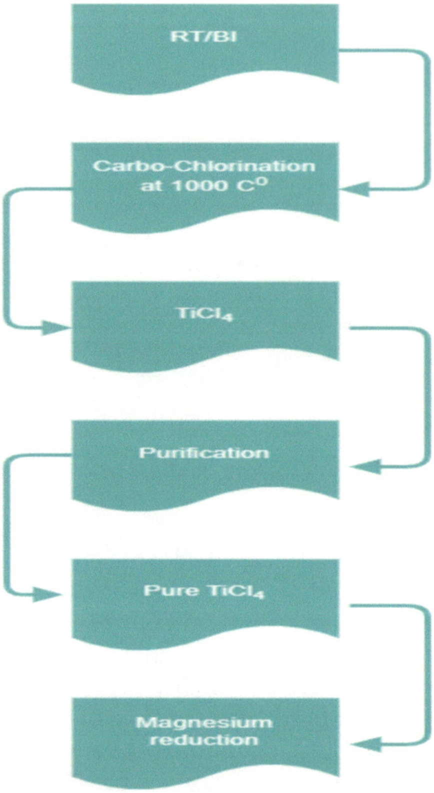

Figure 4.
The Kroll process of titanium extraction.

and post-electrolysis treatment of the sponge [27]. The conventional methods of titanium extraction are widely recognized as established techniques for extracting titanium.

Rutile and beneficiated ilmenite (RT/BI) are both minerals that contain TiO_2 and are used as sources of titanium. Rutile is a naturally occurring mineral composed primarily of titanium dioxide. It has a reddish-brown to black color and is often found in beach sands (**Figure 5**).

Rutile is one of the most common minerals for extracting titanium due to its high titanium content and relatively low impurity levels. RT/BI, on the other hand, refers to ilmenite ore that has undergone a beneficiation process to improve its TiO_2 content and remove impurities. Another important source of titanium is ilmenite that is composed of iron titanium oxide ($FeTiO_3$) as shown in **Figure 6**.

To increase the TiO_2 concentration, the beneficiation technique is applied. The beneficiation process may involve various techniques such as gravity separation, magnetic separation, and flotation. Both rutile and RT/BI are used as feedstocks in the production of TiO_2 pigment, which has a wide range of applications including paints, coatings, plastics, and paper. These minerals are also used as a source of titanium metal through processes such as the Kroll process discussed earlier. Carbo-chlorination at 1000°C is a chemical process used in the extraction of titanium from its ores, specifically rutile (TiO_2) or ilmenite ($FeTiO_3$). In this process, the ore is reacted with carbon usually in the form of coke or coal and chlorine gas (Cl_2) at a high temperature of 1000°C. The carbo-chlorination reaction involves the following steps:

Figure 5.
Rutile rock 5 cm width (coated from Ref. [28]).

Figure 6.
Ilmenite rock includes a high concentration of iron-titanium oxide (coated from Ref. [29]).

Formation of titanium tetrachloride ($TiCl_4$): The carbon reacts with chlorine gas to form carbon monoxide (CO) and carbon tetrachloride (CCl_4). The carbon tetrachloride then reacts with the titanium dioxide in the ore to produce titanium tetrachloride and carbon dioxide (CO_2). Separation of titanium tetrachloride: Titanium tetrachloride is a volatile compound and can be separated from the reaction mixture by condensation, as it has a lower boiling point than other by-products. Overall, carbo-chlorination at 1000°C is an important step in the production of titanium tetrachloride, which serves as a precursor for various titanium-based products, including titanium metal and titanium dioxide pigment. In the Kroll process, magnesium reduction is a key step in the extraction of titanium from its ore, typically rutile or ilmenite. The process involves the following steps: Reduction: The titanium tetrachloride is then reduced using magnesium (Mg) as the reducing agent. The reaction takes place at a high temperature, typically around 900-1000°C, in a closed furnace or reactor [30, 31]. The reduction reaction can be summarized as in the following reaction (Eq. (1)):

$$TiCl_4 + 2Mg \rightarrow Ti + 2MgCl_2 \quad (1)$$

In this reaction, magnesium reacts with titanium tetrachloride to produce titanium metal and magnesium chloride. The reaction is highly exothermic, releasing a significant amount of heat. After the reduction, the mixture is allowed to cool down, and the solid titanium metal is separated from the by-products, mainly magnesium chloride. The separation can be achieved through various methods, such as mechanical processes or selective dissolution in appropriate solvents. The obtained titanium metal may still contain impurities. Therefore, additional purification steps, such as vacuum distillation or other refining processes, may be required to achieve the desired level of purity. The Kroll process, with magnesium reduction as a crucial step, has been the primary method for industrial production of titanium since its development in 1937. It enables the production of high-purity titanium for various applications, including aerospace, chemical processing, and medical implants.

5. Examples of titanium alloys commonly used in aerospace applications

Ti-24Nb-4Zr-8Sn (also known as Ti2448): This alloy refers to compares the microstructure, defects, and mechanical properties of porous structures made from a b-type Ti-24Nb-4Zr-8Sn alloy manufactured using electron beam melting (EBM) and selective laser melting (SLM) techniques. The microstructure of EBM samples consists of $\alpha + \beta$ phases, while SLM samples contain a single β phase due to different powder bed temperatures. The faster cooling rate in SLM results in the formation of fine β dendrites, leading to higher compressive strength (50 ± 0.9 MPa) and lower Young's modulus (0.95 ± 0.05 GPa) compared to EBM parts (45 ± 1.1 MPa and 1.34 ± 0.04 GPa, respectively). Large defects observed in the solid struts are likely caused by tin vaporization, which is more prevalent in SLM due to smaller laser spot size and faster cooling. The number of defects in SLM is approximately 10 times higher than in EBM. These defects have minimal impact on static properties and low-stress fatigue strength but result in reduced and variable fatigue life at high stress levels. Ti-2448, a Ti-24Nb-4Zr-8Sn alloy, exhibits high relative density when using a combination of high-energy density and low laser scanning speed [32, 33]. However, the yield strength of Ti-24Nb-4Zr-8Sn (Ti2448) and pure titanium alloy (Commercially Pure Titanium, Grade 2) was around 275 MPa (40 ksi) and 800-900 MPa (116-130 ksi), respectively. The addition of alloying elements such as Nb, Zr, and Sn in Ti2448 contributes to the solid solution strengthening and precipitation hardening, resulting in enhanced mechanical properties, including higher yield strength. This makes Ti2448 suitable for applications that require higher strength and improved mechanical performance compared to pure titanium alloys [34, 35].

Ti-6Al-4 V (also known as Grade 5 titanium): is renewed for its high strength. The yield strength of grade 5 titanium can vary somewhat depending on factors such as heat treatment, processing condition, and testing method. Different heat treatment process, such as solution treatment and aging, can be employed to enhance the mechanical properties of grade 5 titanium. However, typical values for the yield strength of grade 5 titanium rang from approximately 900-1100 MPa.

Grade 5 titanium offers several benefits based on its mechanical properties including yield strength, tensile strength, and compressive strength. Here are the key advantages of grade 5 titanium:

High strength: Due to high yield strength making it stronger than commercially pure titanium, the alloy's strength allows for the design and manufacture of lightweight components without compromising structural integrity, making it highly suitable for aerospace applications.

Good toughness: Grade 5 titanium also possesses good toughness making it capable of withstanding dynamic loading condition. The dynamic test was tested and concluded the grade 5 titanium alloy sheets allowed to analyze phenomena occurring under dynamic impact loading using a point weight and that revealed the maximum value of absorbed energy is 325 J for three plates. The good behavior of plates maybe of because discrepancy friction between grade 5 titanium layers. From previous study, grade 5 titanium can be tested as a glass reinforcement providing reliable and lightweight solutions for aircraft structures [36]. The specific energy absorption of composite materials can be calculated using:

$$W_{EA} = EA / m\left(KJ / Kg\right)$$

Where EA is energy absorption and m is the mass of composite. Some examples of high-energy absorber materials are shown in **Figure** 7.

Grade 5 titanium is the most widely used titanium alloy in aerospace. It consists of 90% titanium, 6% aluminum, and 4% vanadium. It offers a good balance of strength, toughness, and weldability, making it suitable for various aerospace components, for instance, airframes [36, 37].

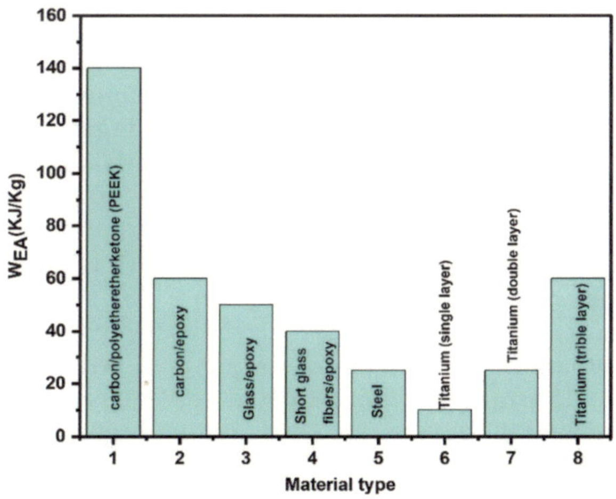

Figure 7.
Specific energy absorption properties depending on material type.

Ti-6Al-2Sn-4Zr-2Mo (also known as Ti 6-2-4-2): This alloy contains titanium, aluminum, tin, zirconium, and molybdenum. It exhibits high strength, excellent fatigue resistance, and good creep resistance at elevated temperatures. It is commonly used in compressor blades, discs, and other high-stress components in gas turbine engines [38]. Ti 6-2-4-2 exhibits high tensile strength typically ranging from 900 MPa to 1100 MPa. The yield strength of Ti 6-2-4-2 alloy is usually around 800 MPa, which is the point at which the material begins to exhibit permanent plastic deformation. The elastic modulus or Young's modulus is approximately 110 GPa This property describes the material's stiffness and its resistance to elastic deformation under applied stress. In addition, this alloy is characterized by resistance to indentation or penetration due to hardness value typically around 300-500 HB (Brinell hardness).

Ti-6Al-2Sn-4Zr-6Mo (also known as Ti 6-2-4-6): The elastic modulus, a measure of material stiffness, for Ti 6-2-4-6 is around 110 GPa (16 x 10^6 psi) [39]. This value indicates its ability to resist deformation under applied loads. It is used in critical aerospace applications such as engine components and structural parts [40].

Ti-5Al-2.5Sn (also known as Ti 5-2.5): This alloy contains titanium, aluminum, and tin. It has good weldability, high strength, and excellent corrosion resistance, particularly in seawater environments. It is commonly used in marine and aerospace applications, including aircraft engine components and offshore structures [41].

Ti 5-2.5 possesses good fatigue strength, allowing it to withstand cyclic loading without failure. The fatigue strength is typically around 450 MPa (65,000 psi) to 515 MPa (75,000 psi), depending on the specific conditions.

Ti-3Al-2.5 V (also known as Grade 9 titanium): This alloy consists of titanium, aluminum, and vanadium. It offers good weldability, high strength, and excellent corrosion resistance. It is primarily used in aircraft hydraulic systems, airframe structures, and engine components [42]. Moreover, Elongation: Grade 9 titanium exhibits good elongation properties, with a typical elongation at break of 15 to 20%. This indicates the alloy's ability to deform before fracture, making it suitable for applications requiring ductility.

Ti-6Al-2Sn-4Zr-2Mo-0.1Si (also known as Ti 6-2-4-2S): This alloy is similar to Ti 6-2-4-2, but with the addition of silicon. It offers improved high-temperature stability and is often used in turbine components, such as blades and disks [12]. The yield strength is around 990 MPa, the ultimate tensile strength (UTS) is around 900-1010 MPa, the ductility El is 8%, and the reduction area is 20% [43].

Ti-6Al-2Sn-2Zr-2Mo-2Cr-0.25Si (also known as α-β grade titanium alloy): This alloy is similar to Ti 6242S but with the addition of chromium. It provides excellent strength, creep resistance, and oxidation resistance, making it suitable for hot section components in gas turbine engines [44]. The yield strength is around 850-910 MPa, the UTS is around 1010-1150 MPa, the elastic modulus is 120 GPa, and the beta-transus temperature (Tβ) is 1020°C [43].

Ti-15 V-3Cr-3Sn-3Al (also known as Ti 15-3-3-3): This alloy consists of titanium, vanadium, chromium, tin, and aluminum. This titanium alloy has high strength, good fatigue resistance, and excellent formability. It is used in applications such as landing gear components, airframe structures, and engine mounts [45]. The yield strength is around 800-1200 MPa, the UTS is around 900-1300 MPa, the elastic modulus is 86-115 GPa, and the beta transus temperature (Tβ) is 795°C [43, 46].

Ti-10 V-2Fe-3Al (also known as Ti 10-2-3): This alloy contains titanium, vanadium, iron, and aluminum. It offers high strength, good weldability, and excellent corrosion resistance. It is commonly used in aerospace applications that require high

DOI: http://dx.doi.org/10.5772/intechopen.1004880

strength and toughness, such as landing gear and structural components [47]. Ti10-2-3 exhibits mechanical properties similar to Ti 15-3-3-3 [43].

Ti-4.5Al-3 V-2Mo-2Fe (also known as Ti 4.5-3-2-2): This alloy contains titanium, aluminum, vanadium, molybdenum, and iron. This alloy is recognized for its excellent strength, high-temperature stability, and corrosion resistance. It is used in aircraft engines and airframe structures, particularly in components exposed to high stress and elevated temperatures [48]. The yield strength is around 900 MPa, the UTS is around 960 MPa, the elastic modulus is 110 GPa, and the beta transus temperature (Tβ) is 900°C [43].

6. Titanium aluminide intermetallic compounds (TiAl)

These intermetallic titanium base alloys are increasingly being used for low-pressure turbine (LPT) blades in aircraft engines such as in Boeing-B777x and A-320neo due to several advantages including high-temperature creep resistance, improved stiffness, and low weight. Recent ongoing research on other Ti alloys such as ATI Titan 23™ and Titan 23™ exhibits improved mechanical properties [12]. Here are the mechanical properties typically associated with TiAl alloys:

Elastic modulus: The elastic modulus of TiAl alloys is relatively low compared to traditional titanium alloys and other metallic materials. It is typically 80 GPa (11.6 x 10^6 psi) [49].

Elongation: TiAl alloys have limited ductility compared to conventional metallic materials. The typical elongation at break is around 1–2%. However, the alloys can exhibit good fracture toughness, which is important for resisting crack propagation [50].

Fatigue strength: TiAl alloys possess good fatigue strength, allowing them to withstand cyclic loading over extended periods. The fatigue strength is 150 MPa, depending on the specific alloy composition and processing conditions [51].

Hardness: The hardness of TiAl alloys can vary depending on the specific composition and heat treatment. It typically 200 HV, with variations based on the desired balance between strength and ductility [52].

Additionally, TiAl alloys offer other advantages in aerospace applications, such as high-temperature resistance, good oxidation and corrosion resistance, and low thermal expansion coefficients.

It is important to note that the mechanical properties of TiAl alloys can vary depending on the specific alloy composition, processing methods, and heat treatments employed. Therefore, it is advisable to refer to the material specifications or consult with the manufacturer for precise mechanical property information for a particular TiAl alloy.

7. Titanium-polymer composite and hybrid materials

Polymer materials play an important role as a modification technique of titanium alloys either as a polymer such as FM-5 or as a polymer composite such as carbon fiber-reinforced plastic. Polymers and polymer composites have very attractive properties combining properties of titanium alloys and the properties of polymers as weight saving, and both are used with metals as adhesives (to join titanium alloy to polymer composites or metals), and adhesion (materials result from a force of connection between titanium alloys and polymers). Both adhesion and adhesive must operate in high-temperature environments and maintain their mechanical

Figure 8.
Chemical interaction mechanism of adhesively bonded joints [53].

properties to operate successfully for aerospace applications. Titanium/polymer adhesion can be synthesized by the polymerization of monomers on the titanium surface to form adhesion bonding between these components. It should be noted that two mechanisms are possible to elucidate the bonded joints: chemical interactions and mechanical interlocking as shown in **Figures 8** and **9**, respectively [53]. Titanium alloys/polymers-based hybrid materials have many advantages such as protection and weight reduction [54]. Titanium alloy hybrid materials can be explained by the joint between Ti and carbon-fiber-reinforced polymer (Ti-CFRP).

An example of the unique applications of using polymers with titanium alloys is to improve the antistatic properties of metal-matrix composites as well as corrosion resistance [55]. Composite materials consist of a matrix and a filler resulting in a composite that combines both matrix and filler properties, while hybrid materials consist of mixed components resulting in a material with individual properties. There are titanium-polymer composites and hybrids used in aerospace applications that combine the properties of titanium alloys with the benefits of polymers. In addition, these titanium-polymer composites are designed to provide a balance between the desirable properties of titanium alloys and polymers, offering improved performance and versatility in aerospace applications. These composites offer lightweight materials with improved mechanical properties and enhanced resistance to fatigue and impact.

Figure 9.
Mechanical interlocking mechanism of adhesively bonded joints [53].

	Modulus of elasticity (GPa)	Ultimate Tensile Strength (MPa)	Elongation at break (%)	Hardness (Rockwell M)
PEEK	2.14-7.58	50.3-265	1.5-110	85-109
PEI	0.00280-56.0	1.00-600	0.50-110	85-118
PPS	1.56-17.2	27.6-155	0.35-80	90-100
PEI-CF	4.50-56.0	124-650	0.50-8.0	90-105

Table 2.
Some mechanical properties of PEEK, PEI, PPS, and PEI-CF [60–64].

Titanium/polyetheretherketone (PEEK) composite: PEEK is a high-performance polymer that is often reinforced with titanium fibers or particles. The combination of titanium and PEEK results in a composite material with increased strength, stiffness, and heat resistance. It is used in aerospace applications such as brackets, connectors, and structural components [56].

Titanium/polyetherimide (PEI) composite: PEI is a thermoplastic polymer that can be reinforced with titanium fibers or particles. The resulting composite exhibits excellent mechanical strength, chemical resistance, and high-temperature performance. It is used in aerospace applications that require lightweight and durable components, such as aircraft interior parts and engine components [57].

Titanium/polyphenylene sulfide (PPS) composite: PPS is a high-performance thermoplastic polymer that can be combined with titanium fibers or particles. The composite offers good mechanical strength, chemical resistance, and dimensional stability. It finds applications in aerospace where lightweight, strong, and corrosion-resistant materials are needed, such as aircraft structural components and engine parts [58].

Titanium/PEI-carbon fiber (PEI-CF) composite: Carbon fiber-reinforced PEI, also known as PEI-CF, can be further reinforced with titanium fibers or particles. This composite exhibits high strength, stiffness, and excellent thermal stability. It is used in materials for aerospace applications that require lightweight and high-performance characteristics, such as aircraft panels, wing structures, and rotor blades [59].

The composition of titanium alloys with polymers provides materials with unique mechanical properties. Titanium enhances the overall strength, stiffness, and density. A specific property of titanium alloys can be modified by composition with a polymer with this property. For example, PEEK is chemical-resistant, biocompatible, flexible and exhibits high tensile strength. PEI in contrast has a higher range modulus of elasticity than PEEK. The UTS of the PEI-CF composite is higher than the pure polymers as shown in **Table 2**. In addition, the mechanical properties of titanium/polymer composite materials can be manipulated by many factors such as the ratio between the titanium alloy and the selected polymer, the bond mechanism, the method of distribution, the particle size, and the effect on the interfacial strength of the titanium/polymer composite joint.

8. Factors affecting the modification of titanium alloys

Surface pretreatments of many titanium alloys have been studied to evaluate the effect on bond strength between the adhesive and the adherend's surface and to avoid cohesive failures. According to a previous study, surface pretreatments on adhesive properties were measured by the contact angle between titanium and carbon fiber-reinforced plastic (CFRP). For instance, treatments of both substrate surfaces (titanium and the CFRP composite) include UV/ozone method to increase the surfaces' free energy and create wetting characteristics [65].

Other factors affect the modification of titanium alloys as the following [17, 54]:

- Ultrasonic welding for Ti-Nylon-6 CFRTP lap joints
- Acid-picking (etching) of Ti alloy surfaces
- Anodization
- Vibratory shot peening
- Brazing process
- Heating pretreatment
- Alkaline degreasing
- Priming

Another surface treatment can be achieved by plasma polymerization on a titanium alloy's surface which results in stable corrosion protection and strengthens the adhesive bonding. An example of this method was described using argon plasma polymerization of Si-films on the Ti6AL4V substrate to study the effect of the deposition pressure on surface properties such as layer thickness and surface roughness [66].

The plasma treatment significantly affected the surface properties and bond strength of these adhesives. The durability and mechanical properties of adhesive joints between Ti6AL4V with polyimides such as FM-5 and PETI-5 were investigated as promising materials for high-speed aircraft applications. The presence of high thermal characteristics of polyimides promotes these types of adhesives [67]. Polyether ether ketone (PEEK) is another class of polymers that have been used with plasma-nitrided titanium sheets owing to unique thermal and mechanical properties that meet the requirements of aerospace applications [68].

Other unique properties of certain titanium alloys are shape memory and pseudoelasticity behaviors (the ability to recover the original shape after being changed by temperature and pressure). The properties of interfacial adhesion of Ti-alloys and the other metal or polymer composites can be enhanced. Ti-Ni alloy is an example of these materials that have been used as smart materials and active materials in the aerospace area [69].

Recently, a study conducted a synthesis of a strong 3-dimensional (3D) interface configuration of matrix composite of the graphene nanoparticles-reinforced titanium alloy (Ti6AL4V) with the adherent TiB_2 resulting in excellent strength and ductility [70].

9. Conclusion

In conclusion, this chapter discussed the mechanical properties and performance of titanium-based alloys in aerospace applications. The historical background and development of titanium alloys for aerospace were presented, highlighting the significance and demand for these alloys in the industry. The titanium alloys based on equilibrium at room temperature were explored, showcasing their diverse properties and applications.

The chapter also provides an overview of the titanium extraction process, shedding light on the complex manufacturing methods involved in producing these alloys. Several examples of titanium alloys commonly used in aerospace applications were mentioned, emphasizing their suitability for high-strength and lightweight structures.

Furthermore, the discussion extended to titanium aluminide intermetallic compounds and their potential benefits in aerospace engineering. The unique properties of titanium aluminide intermetallic alloys, such as high temperature resistance and excellence strength-to-weight ratio, make them promising candidates for advanced aerospace applications.

Lastly, the chapter touched upon the emerging field of titanium-polymer composites and hybrid materials. These innovative materials combine the desirable properties of titanium alloys with the advantages of polymer matrices, resulting in enhance performance and versatility.

In summary, the study of mechanical properties and performance of titanium-based alloys in aerospace applications is crucial for advancing the field of aerospace engineering. The continuous development utilization of these alloys, along with the exploration of new materials and manufacturing techniques, will contribute to the improvement of aircraft performance, fuel efficiency, and overall safety in the aerospace industry.

Acknowledgements

The authors would like to thank Dean of Scientific Research, King Khalid University, Abha, Saudi Arabia, for their encouragement to accomplish this chapter.

Conflict of interest

The authors declare no conflict of interest.

Thanks

This work is dedicated to the authors' parents and to those who inspired it.

Acronyms and abbreviations

OEW	operation empty weight
α	alpha
β	beta
Ti	titanium
HCP	hexagonal close-packed
BCC	centered cubic crystal
TiO_2	titanium dioxide

°C	celsius
g/cm^3	gram/centimeter
GPa	gigapascals
MPa	megapascals
W/mK	Watts per meter kelvin
$TiCl_4$	titanium tetra chloride
RT/BI	rutile and beneficiated ilmenite
$FeTiO_3$	iron titanium oxide
Cl_2	chlorine
CO	carbon monoxide
CCl_4	carbon tetrachloride
CO_2	carbon dioxide
Mg	magnesium
EBM	electron beam melting
SLM	selective laser melting
TiAl	titanium aluminide
LPT	low-pressure turbine
Ti-CFRP	Ti and carbon-fiber-reinforced polymer
PEEK	polyetheretherketone
PEI	polyetherimide
PPS	polyphenylene sulfide
PEI-CF	PEI-carbon fiber
CFRP	carbon fiber-reinforced plastic

Author details

Amal Mohammed Abdulrahman[1], Abdul Mohshen Sharif Ullah Siddique[1*] and Rayanah Barnawi[2]

1 Department of Chemistry, College of Science, King Khalid University (KKU), Abha, Kingdom of Saudi Arabia (KSA)

2 Microgravity Research Department, Exploration and Science Sector, Saudi Space Agency, Riyadh, Kingdom of Saudi Arabia (KSA)

*Address all correspondence to: abdulmohshen@gmail.com

References

[1] Whittaker M. Titanium alloys. Metals. 2015;**5**:1437-1439. DOI: 10.3390/met5031437

[2] Prasad Y, Seshacharyulu T. Processing maps for hot working of titanium alloys. Materials Science and Engineering A. 1998;**243**:82-88. DOI: 10.1016/S0921-5093(97)00782-X

[3] Brewer W, Bird R, Wallace T. Titanium alloys and processing for high speed aircraft. Materials Science and Engineering A. 1998;**243**:299-304. DOI: 10.1016/S0921-5093(97)00818-6

[4] Chunxiang C, BaoMin H, Lichen Z, Shuangjin L. Titanium alloy production technology, market prospects and industry development. Materials & Design. 2011;**32**:1684-1691. DOI: 10.1016/j.matdes.2010.09.011

[5] Kothari K, Radhakrishnan R, Wereley N. Advances in gamma titanium aluminides and their manufacturing techniques. Progress in Aerospace Sciences. 2012;**55**:1-16. DOI: 10.1016/j.paerosci.2012.04.001

[6] Wartman F, Baker D, Nettle J, Homme V. Some observations on the Kroll process for titanium. Journal of the Electrochemical Society. 1954;**101**:507-513. DOI: 10.1149/1.2781146

[7] Kroll W. How commercial titanium and zirconium were born. Journal of the Franklin Institute-Engineering and Applied Mathematics. 1955;**260**:169-192. DOI: 10.1016/0016-0032(55)90727-4

[8] Froes F, Imam M. Cost affordable developments in titanium technology and applications. Key Engineering Materials. 2010;**436**:1-11. DOI: 10.4028/www.scientific.net/kem.436.1

[9] Farthing T. Introducing a new material. The story of titanium. Proceedings of the Institution of Mechanical Engineers. 1977;**191**:59-73. DOI: 10.1243/PIME_PROC_1977_191_012_02

[10] Subramanya R, Sathyanarayana P, Kumar S, Thangarasu V, Manjunatha K. High speed machining of titanium Ti 6Al4V alloy components: Study and optimisation of cutting parameters using RSM. Advances in Materials and Processing Technologies. 2022;**8**:277-290. DOI: 10.1080/2374068X.2020.1806684

[11] Rawal S, Brantley J, Karabudak N. Additive manufacturing of Ti-6Al-4V alloy components for spacecraft applications. In: Sixth International Conference on Recent Advances in Space Technologies (RAST'6) 12-14 June 2013. Istanbul: IEEE; 2013. pp. 5-11

[12] Williams J, Boyer R. Opportunities and issues in the application of titanium alloys for aerospace components. Metals. 2020;**10**:705. DOI: 10.3390/met10060705

[13] Petrescu RVV. About supersonic flight and mach 3 flying. American Journal of Engineering and Applied Sciences. 2020;**13**:451-476. DOI: 10.3844/ajeassp.2020.451.476

[14] Nyamekye P, Golroudbary S, Piili H, Luukka P, Kraslawski A. Impact of additive manufacturing on titanium supply chain: Case of titanium alloys in automotive and aerospace industries. Advances in Industrial and Manufacturing Engineering. 2023;**6**:1-18. DOI: 10.1016/j.aime.2023.100112

[15] Joshi S, Sheikh A. 3D printing in aerospace and its long-term

sustainability. In: First International Conference on Progress in Additive Manufacturing (pro-AM 2014). Virtual and Physical Prototyping United Kingdom; 2015. pp. 175-185

[16] Chattoraj I. Stress corrosion cracking (SCC) and hydrogen-assisted cracking in titanium alloys. In: Stress Corrosion Cracking. Cambridge: Woodhead Publishing; 2011. pp. 381-408. DOI: 10.1533/9780857093769.3.381

[17] Rudawska A, Zaleski K, Miturska I, Skoczylas A. Effect of the application of different surface treatment methods on the strength of titanium alloy sheet adhesive lap joints. Boston: Materials. 2019;**12**:4173-4189. DOI: 10.3390/ma12244173

[18] Veiga C, Davim J, Loureiro A. Properties and applications of titanium alloys: A brief review. Advances in Materials Science. 2012;**12**:14-34

[19] Illarionov G, Nezhdanov A, Stepanov S, Muller-Kamskii G, Popov A. Structure, phase composition, and mechanical properties of biocompatible titanium alloys of different types. The Physics of Metals and Metallography. 2020;**121**:367-373

[20] Peters M, Kumpfert J, Ward CH, Leyens C. Titanium alloys for aerospace applications. Advanced Engineering Materials. 2003;**5**:419-427. DOI: 10.1002/adem.200310095

[21] Cui C, BaoMin H, Lichen Z, Shuangjin L. Titanium alloy production technology, market prospects and industry development. Materials and Design. 2011;**32**:1684-1691. DOI: 10.1016/j.matdes.2010.09.011

[22] Kuttolamadom M, Jones J, Mears L, Kurfess T. Investigation of the machining of titanium components for lightweight vehicles. SAE Technical Paper. 2010;**01**:0022. DOI: 10.4271/2010-01-0022

[23] Titanium Ti [Internet]. Available from: https://www.matweb.com [Accessed: December 15, 2023]

[24] Singh P, Pungotra H, Kalsi N. On the characteristics of titanium alloys for the aircraft applications. Materials Today: Proceedings. 2017;**4**:8971-8982. DOI: 10.1016/j.matpr.2017.07.249

[25] Boyer R. Titanium for aerospace: Rationale and applications. Advanced Performance Materials. 1995;**2**:349-368

[26] Kazemi ME, Shanmugam L, Yang L, Yang J. A review on the hybrid titanium composite laminates (HTCLs) with focuses on surface treatments, fabrications, and mechanical properties. Composites Part A: Applied Science and Manufacturing. 2020;**128**:105679. DOI: 10.1016/j.compositesa.2019.105679

[27] Nagesh C, Ramachandran S, Subramanyam R. Methods of titanium sponge production. Transactions of the Indian Institute of Metals. 2008;**61**:341-348

[28] Rutil [Internet]. Available from: https://www.sandatlas.org/rutile [Accessed: January 20, 2024]

[29] Ilimenite [Internet]. Available from: https://geology.com/minerals/ilmenite.shtml. [Accessed: January 20, 2024]

[30] Zhang Z, Matsuura H, Tsukihashi F. Chapter 4.4 - Processes for Recycling. Boston: Elsevier; 2014. pp. 1507-1561. DOI: 10.1016/B978-0-08-096988-6.00036-5

[31] Gerdemann S. Titanium process technologies. Advanced Materials & Processes. 2001;**159**:41-43

[32] Mosallanejad M, Abdi A, Karpasand F, Nassiri N, Iuliano L, Saboori A. Additive manufacturing of titanium alloys: Processability, properties, and applications. Advanced Engineering Materials. 2023;**25**:2301122. DOI: 10.1002/adem.202301122

[33] Liu Y, Li S, Wang H, Hou W, Hao Y, Yang R, et al. Microstructure, defects and mechanical behavior of beta-type titanium porous structures manufactured by electron beam melting and selective laser melting. Acta Materialia. 2016;**113**:56-67. DOI: 10.1016/j.actamat.2016.04.029

[34] Totten G, Xie L, Funatani K. Handbook of Mechanical Alloy Design. Boca Raton: Taylor & Francis; 2003. 700 p. DOI: 10.1201/9780203913307

[35] Nune K, Misra K, Li S, Hao Y, Yang R. Cellular response of osteoblasts to low modulus Ti-24Nb-4Zr-8Sn alloy mesh structure. Journal of Biomedical Materials Research Part A. 2017;**105**(3):859-870. DOI: 10.1002/jbm.a.35963

[36] Głowacki D, Moćko W, Marczak M, Głowacka A, Kraśkiewicz C. Energy absorbing properties analysis of layers structure of titanium alloy Ti6Al4V during dynamic impact loading tests. Materials. 2021;**14**(23):7209. DOI: 10.3390/ma14237209

[37] Szusta J, Tüzün N, Karakaş Ö. Monotonic mechanical properties of titanium grade 5 (6Al-4V) welds made by microplasma. Theoretical and Applied Fracture Mechanics. 2019;**100**:27-38. DOI: 10.1016/j.tafmec.2018.12.009

[38] Jaquez-Muñoz J, Gaona-Tiburcio C, Lira-Martinez A, Zambrano-Robledo P, Maldonado-BandalaE,Samaniego-GamezO, et al. Susceptibility to pitting corrosion of Ti-CP2, Ti-6Al-2Sn-4Zr-2Mo, and Ti-6Al-4V alloys for aeronautical applications. Metals. 2021;**11**:1002. DOI: 10.3390/met11071002

[39] Lütjering G, Williams J, Gysler A. Microstructure and Mechanical Properties of Titanium Alloys. Vol. 2. Singapore: World Scientific; 2000. pp. 1-77

[40] Davies P, Johal A, Davies H, Marchisio S. Powder interlayer bonding of titanium alloys: Ti-6Al-2Sn-4Zr-6Mo and Ti-6Al-4V P. International Journal of Advanced Manufacturing Technology. 2019;**103**:441-452. DOI: 10.1007/s00170-019-03445-3

[41] Khan A, Rahman M. Surface characteristics of Ti-5Al-2.5Sn in electrical discharge machining using negative polarity of electrode. International Journal of Advanced Manufacturing Technology. 2017;**92**:1-12. DOI: 10.1007/s00170-017-0028-4

[42] Ahmed Y, Sahari K, Ishak M, Khidhi B. Titanium and its alloy. International Journal of Science and Research (IJSR). 2012;**3**:358

[43] Ahmed YM, Sahari K, Ishak M, Khidhir B. Titanium and its alloy. International Journal of Science and Research (IJSR). 2014;**3**:1351-1361

[44] Baloyi P. Reinforcement of mild steel AISI 1008 –MIG welds using Titanium powder [thesis]. Johannesburg: University of Johannesburg; 2021

[45] Yumak N, Aslantas K. Effect of heat treatment procedure on mechanical properties of Ti-15V-3Al-3Sn-3Cr metastable b titanium alloy. MEPEG. 2021;**30**:1066-1074. DOI: 10.1007/s11665-020-05445-x

[46] Pushp P, Dasharath SM, Arati C. Classification and applications

of titanium and its alloys. Materials Today: Proceedings. 2022;**54**:537-542

[47] Duerig T, Allison J, Williams J. Microstructural influences on fatigue crack propagation in Ti-10V-2Fe-3Al. Metallurgical Transactions A. 1985;**16**:739-751. DOI: 10.1007/BF02814825

[48] Takeda J, Niinomi M, Akahori T. 605 fretting fatigue characteristics and microstructure of Ti-4.5 Al-3V-2Mo-2Fe for aircraft applications. In: The Proceedings of the JSME Materials and Processing Conference (M&P); 15-18 October 2002. Japan: JSME; 2002. pp. 452-457

[49] Cox BN, James MR, Marshal DB, et al. Determination of residual stresses in thin sheet titanium aluminide composite. Metallurgical and Materials Transactions A (Physical Metallurgy and Materials Science). 1990;**21A**:2701-2707

[50] Baeslack I, Mascorella T, Kelly T. Weldability of a titanium aluminide. Welding Journal. 1989;**68**:12

[51] Hao Y, Jibin Z, Tianran W. Research on a different method to reach the saturate limit of titanium aluminide alloy surface mechanical and fatigue properties by laser shock process. Optik. 2019;**193**:162989. DOI: 10.1016/j.ijleo.2019.162989

[52] Sinan A, Yavuz K. Formation of Ti-aluminides on commercially pure Ti via the hot-dipping aluminizing process. Transactions of the Indian Institute of Metals. 2020;**73**:1065-1072. DOI: 10.1007/s12666-020-01948-2

[53] Lobbecke M, Haubrich J. Bonding and aging mechanisms of polymers on titanium alloys. In: European Congress and Exhibition on Advanced Materials and Processes 1-5 September 2019. Stockholm: Euromat; 2019

[54] Faudree M, Uchida H, Kimura H, Kaneko S, Salvia M, Nishi Y. Advances in titanium/polymer hybrid joints by carbon fiber plug insert: Current status and review. Materials. 2022;**15**:3220-3239. DOI: 10.3390/ma15093220

[55] Yadav R, Tirumali M, Wang X, Naebe M, Kandasubramanian K. Polymer composite for antistatic application in aerospace. Defence Technology. 2020;**16**:107-118. DOI: 10.1016/j.dt.2019.04.008

[56] Su Y, Rooij M, Grouve W, Akkerman R. The effect of titanium surface treatment on the interfacial strength of titanium–thermoplastic composite joints. International Journal of Adhesion and Adhesives. 2017;**72**:98-108. DOI: 10.1016/j.ijadhadh.2016.10.007

[57] Catauro M, Raucci M, Ausanio G, Ambrosio L. Sol-gel synthesis, characterization and bioactivity of poly(ether-imide)/TiO_2 hybrid materials. Journal of Applied Biomaterials & Biomechanics. 2007;**5**:41-48. DOI: 10.1177/228080000700500106

[58] Du M, Dong W, Dong L, Li X, Wang L. Nanostructure modification of titanium alloy to achieve ultra-high interfacial bond strength between titanium alloy and polyphenylene sulfide. Journal of Materials Research and Technology. 2023;**26**:3383-3394. DOI: 10.1016/j.jmrt.2023.08.116

[59] Xiong H, Gao T, Li K, Liu Y, Ma Y, Liu J, et al. A polymer-oriented self-assembly strategy toward mesoporous metal oxides with ultrahigh surface areas. Advancement of Science. 2019;**6**:1801543-1801552. DOI: 10.1002/advs.201801543

[60] Liao C, Li Y, Tjong SC. Polyetheretherketone and its composites for bone replacement and regeneration. Polymers (Basel). 2020;**12**(12):2858. DOI: 10.3390/polym12122858

[61] Polyetheretherketone (PEEK) [Internet]. Available from: https://www.matweb.com [Accessed: February 10, 2024]

[62] Polyetherimide (PEI) [Internet]. Available from: https://www.matweb.com [Accessed: February 10, 2024]

[63] Polyphenylene Sulfide (PPS) [Internet]. Available from: https://www.matweb.com [Accessed: February 10, 2024]

[64] Polyetherimide (PEI). Carbon fiber filled [Internet]. Available from: https://www.matweb.com [Accessed: February 10, 2024]

[65] Wang W, Poulis J, Teixeira De Freitas S, Zarouchas D. Surface pretreatments on CFRP and titanium for manufacturing adhesively bonded bi-material joints. In: 18th European Conference on Composite Materials (ECCM18), 24-28 June 2018. Athens, Greece: ECCM; 2020. pp. 1-8

[66] Pereira M, Baldin E, Antonini L, Bernardi F, Oliveira L, Maurmann N, et al. TEOS thin films obtained by plasma polymerization on Ti6Al4V alloys: Influence of the deposition pressure on surface properties and cellular response. Applied Surface Science. 2021;**5**:1-19. DOI: 10.1016/j.apsadv.2021.100123

[67] Xu S, Guo S, Dillard D. Evaluation of the long-term durability of high-performance polyimide adhesives for bonding titanium. The Journal of Adhesion. 2004;**80**:1153. DOI: 10.1080/00218460490884330

[68] Ahmed S, Bhowmik S. Interpenetrating polymer network adhesive bonding of PEEK to titanium for aerospace application. Journal of Polymer Engineering. 2019;**39**(1):1-9. DOI: 10.1515/polyeng-2018-0148

[69] Costanza G, Tata M. Shape memory alloys for aerospace, recent developments, and new applications: A short review. Materials. 2020;**13**:1856. DOI: 10.3390/ma13081856

[70] Ren W, Zhang W, Zhou S, Zhou Q, Wei J, Wu P, et al. Achieving high strength-ductility in TiBw-GNPs/Ti6Al4V composites via 3D interface configuration. Journal of Alloys and Compounds. 2024;**971**:172809. DOI: 10.2139/ssrn.4369211

Chapter 3

Galling-Free Forming of Titanium and Titanium Alloys Using Carbon-Supersaturated Tool Steel Dies

Tatsuhiko Aizawa, Kenji Fuchiwaki, Takeshi Kihara and Tomomi Shiratori

Abstract

Carbon supersaturation (CS) process was developed to prepare the CS-tool steel dies with massive carbon solute content toward the galling-free metal forming. The impinged carbon solutes diffused and agglomerated onto the hot spots at the die-work interface by stress gradient during the metal forming. This in situ formed free-carbon thin film worked as a tribofilm to reduce the friction and adhesive wear on the die-work interface. Titanium and titanium alloys were selected as a work material common to forging, near-net forming and fine blanking processes. The ball-on-disc method was employed to demonstrate the significant reduction of friction coefficient by CS-tool steels against the pure titanium ball. Upsetting process was used to describe the galling-free forging behavior even under the higher reduction of thickness than 50%. Pin-forming process was utilized to prove that taller pins than designed target were extruded and their height was preserved even with increasing the number of strokes. Fine blanking process was used to describe the integrity of CS-punch with higher grade of titanium gears. The in situ solid lubrication by formation of free-carbon tribofilm was discussed in each metal forming. In particular, the initial learning trial was proposed to shorten the incubation time for the free-carbon film coverage onto the hot spots.

Keywords: galling-free forming, titanium and titanium alloys, In situ solid lubrication, free-carbon tribofilm, upsetting, pin forming, fine blanking, high qualification of products

1. Introduction

Titanium and titanium alloys have been widely utilized in every industry and medical application [1]. During their metal forming, their fresh surfaces come into contact with the die surface, resulting in severe adhesion wear or galling damage to dies [2]. Hence, the lubrication is indispensable for their metal forming in practice. Manufacturing of eye-glass flames and medical parts disliked the contamination of lubricating oils; solid lubrication provides an only way to control the contact interface condition between the work and the dies [3, 4]. After [5], the bulk ceramic dies suffered from severe adhesion of pure titanium plate to lower the limit of deep drawability. The ceramic coatings, such as TiN, TiCN or TiAlN and DLC (Diamond-Like Coating), were not free from galling, so that higher friction coefficient than 0.2 was attained in the ball-on-disc testing [6]. As discussed in [7], the galling often occurs during hot forming processes even at the presence of solid lubricants and ceramic coatings under relatively small plastic strains. Among various solid lubrications, the in situ solid lubrication methods become only a solution to reduce the friction and wear on the die-work interface and to sustain low-frictional and wearing state although the steps in forming.

Figure 1 compares three methods for in situ solid lubrication [8]. The chlorine-implanted TiN and TiCN coatings had anti-galling capacity due to in situ formation of deformable titanium oxide (TiO_x for $1 < x < 2$) film on the contact interface [9–11], as depicted in **Figure 1a**. Tetragonal DLC (t-a:C) coating, working in the tailored polymer lubricant, hindered the direct contact of work materials with DLC coating by in situ synthesis of polymer layer to attain the superlubrication conditions [12], as illustrated in **Figure 1b**. Using the carbon supersaturated ceramic, stainless steel and tool steel dies, the isolated carbon film from CS-dies was in situ formed on the contact die-work interface to lower the friction and wear [13–18].

Among three solid lubricating mechanisms, the in situ formed carbon tribofilm works in practical metal forming, where the tool steel dies are used for dry and semi-dry forming of titanium and titanium alloys. This in situ formation of free-carbon tribofilms, in **Figure 1c**, is further explained in **Figure 2**.

The lower temperature plasma carburizing process [13–15] was utilized for massive carbon supersaturation in tool steel or stainless steel dies. The plasma carburized dies at 673 K for 14.4 ks had a CS-layer with the thickness of 40 μm and the average carbon solute content of 3 mass%. Most carbon solutes occupy the octahedral vacancy sites in the iron lattice cells to induce the lattice expansion. The lattices in CS-layer strained in elasticity, while the neighboring cells to CS-layer plastically strained to compensate for the strain incompatibility across the carburizing front

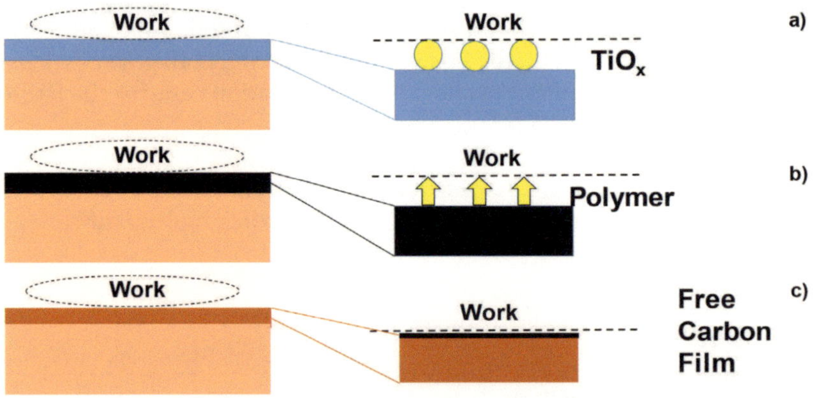

Figure 1.
Typical in situ solid lubrication mechanisms during the metal forming even in dry or in MQL (minimum quantity lubrication). a) In situ formation of deformable titanium oxide film (TiO_x for $1 < x < 2$) on the titanium base coating to work interface, b) in situ formation of monolayer polymer film onto the t-a:C coating and c) in situ formation of free-carbon film onto the CS-tool to work interface.

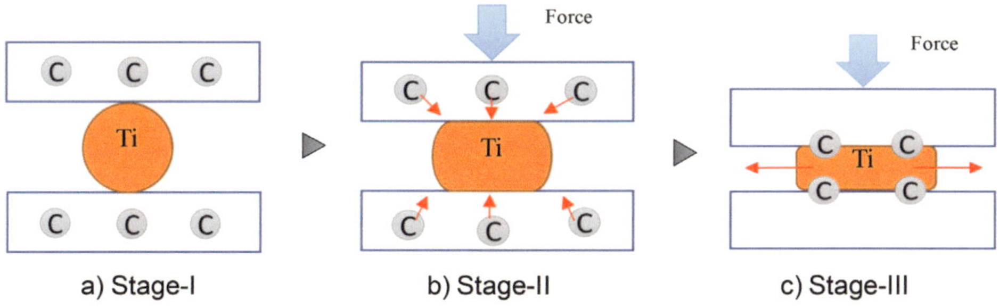

Figure 2.
Illustration of this in situ solid lubrication process. a) Stage-I: massive carbon supersaturation to tool steel dies, b) stage-II: diffusion and agglomeration of isolated free-carbon solutes from CS-dies and c) stage-III: formation of free-carbon tribofilm onto the highly stressed hot spots at the CS-tool to titanium work interface.

end. Due to this plastic straining, the original die microstructure was nano-structured to have the specific orientation to (111) or easy-to-deformation axis [19, 20]. In local, the carbon solute occupation had disturbance by its chemical affinity among iron/nickel and chromium, the constituent compositions in the stainless steels or tool steels. Then, this nanostructure had two-cluster decomposition system, where the carbon-rich cluster was neighboring to the carbon-poor cluster in the nano-structured CS-layer. The carbon solute was easy to isolate from its occupation sites and to jump to next sites through this fine cluster boundary network by externally applied driving force in **Figure 2a** [21–23].

On the true contact state between the fresh work material and the CS-dies during metal forming, high traction was applied to the hot spots at the contact interface. This high stress gradient induced the carbon solute jumping diffusion to the hot spots. These attracted free-carbon solutes by stress gradients, agglomerated on the hot spots and formed a free-carbon tribofilm to hinder the direct contact of work with CS-die surfaces in **Figure 2b**. During the metal forming, this tribofilm gradually grew and became a stable tribofilm on the CS-die to work interface in **Figure 2c**. As demonstrated in [24–26], this in situ solid lubrication mechanism via CS-tools and CS-dies played a significant role to reduce the friction, the wearing and the work hardening in forming and shaping of titanium and titanium alloys.

In the present chapter, this in situ solid lubrication process was experimentally described and analyzed to demonstrate its effectiveness to prevent the tools and dies from severe galling behavior in metal forming. First, the ball-on-disc method is employed to explain the gradual formation of the free-carbon tribofilm on the track of CS-disc with an increase in sliding distance. Low friction and wearing state are preserved in longer sliding distance against the pure titanium ball. In second, three metal forming processes are employed to actually prove that low friction and wearing state are also sustained by this in situ solid lubrication—e.g., forging, pin forming and fine blanking. Higher reduction of thickness than 50% is attained by a single-step upsetting process with low friction, low adhesive wear and low work hardening. A titanium round bar is continuously free forged to have a triangular cross-section beam within a single stroke. Taller pins than designed target are formed into the forged titanium frame in every stroke for mass production. This tall pin height is preserved although the extrusion steps in eye-glass frame manufacturing. A titanium gear with high dimensional accuracy and low surface roughness is continuously yielded by fine blanking without severe galling of titanium. High product qualification was also proved by dimensional comparison of products with and without carbon supersaturation. The features of this approach are discussed to improve the metal forming performance especially in its early stage and to extend this technology to other industrial and medical fields.

2. Methods and materials

The plasma carburizing system is explained with comments on the processing conditions for massive carbon supersaturation to tool steel dies. The ball-on-disc (BOD) testing and three metal forming procedures are stated to experimentally demonstrate the effectiveness of this in situ solid lubrication mechanism as a unique process tribological method for metal forming of titanium and titanium alloys.

2.1 Plasma carburizing process

RF (Radio-Frequency) and DC (Direct Current) generators were used in this plasma carburizing process to ignite and preserve the plasma-immersion conditions. This system consisted of RF- and DC-plasma generators, the driving gas suppliers, the evacuation unit and the EOS (Emission Optical Spectroscopy) unit for plasma diagnosis as illustrated in **Figure 3a**. The hollow cathode device was employed to keep high carbon-ion and CH-radical densities after this diagnosis and to deepen the carbon supersaturated layer thickness [27–29]. The dipole electrodes were utilized to ignite the RF plasma, while the DC plasma was also induced by applying the bias voltage to the bottom of hollow cathode. The RF-voltage and the DC-bias were constant by +210 V and – 500 V, respectively. The carburized layer thickness was estimated to be 40 μm in case of the plasma carburizing at 673 K for 18.0 ks by 50 Pa.

The following procedure was employed in practical operation for plasma carburizing. After evacuation down to the base pressure of 0.01 Pa, the argon gas was introduced to a chamber at RT to clean the punch and die surfaces in **Figure 3b** via presputtering. After increasing the process temperature up to 673 K under the argon atmosphere, the hydrogen gas was also introduced into the argon gas with the flow rate of 160 mL/min for argon and 20 mL/min, respectively. The total pressure was constant by 50 Pa. After presputtering by the DC-plasmas for 1.8 ks, the methane gas

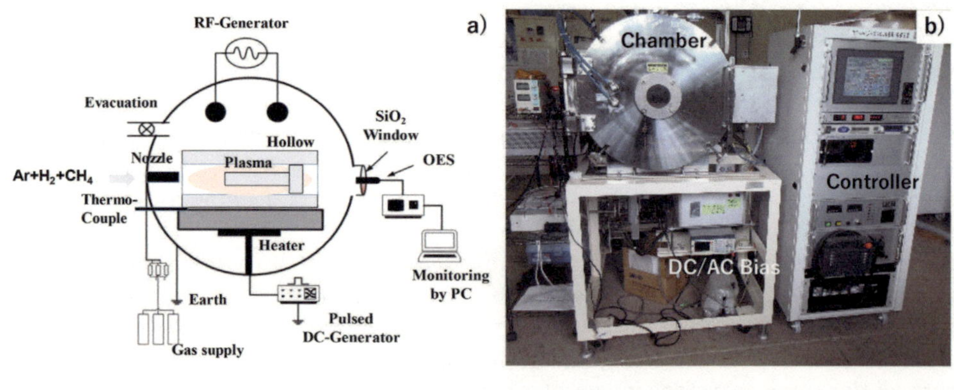

Figure 3.
Low temperature plasma carburizing system. a) Its schematic view of the ignition of plasmas in the hollow cathode and b) overview of the whole system.

was introduced as a carbon source into argon and hydrogen mixture gas by the flow rate of 20 mL/min. At the specified duration of 18.0 ks, the specimen was cooled down in the chamber under the nitrogen atmosphere before evacuation down to atmospheric pressure. The processing temperature was in situ monitored by the thermocouple, which was embedded into the base plate below the hollow cathode device.

2.2 Ball-on-disc testing

BOD tribotesting system (Tribometer; CSM, Switzerland) in the rotational mode was utilized to describe the frictional behavior between the CS-SKD11 disc and the pure titanium ball in **Figure 4**. The SKD11 disc with the diameter of 60 mm and the thickness of 5 mm was prepared for carbon supersaturation. The pure titanium ball with the diameter of 6 mm was used as a counter material. The friction coefficient (μ) was calculated by $\mu = F/W$ for the measured frictional force, F, and the applied normal load, W. The sliding velocity, V, was varied together with W to describe the variation of the friction coefficient with increasing the sliding distance, L, up to L = 500 m.

2.3 Metal forming procedure

CNC (Computer Numerical Control)-stamping system (Hoden Seimitsu; Kanagawa, Japan) was employed for upsetting experiments with the use of two different CS-die pairs, as depicted in **Figure 5**.

CS-SKD11 punch and die were, respectively, fixed into the upper and lower cassette die sets. Those two die sets were cemented to upper and lower bolsters of CNC-stamping system with the stroke control capability. The stroke was controlled to move down the upper bolster to the specified position for reduction of thickness in work materials. The loading speed was constant by 10 mm/s.

Figure 6 depicts the upsetting process to infiltrate the pure titanium work into the die cavities. Since a fresh titanium work contacts the cavity walls and die edges, this process has a risk of severe galling with high friction. This high wearing and friction results in shortage of pin height; the accumulated adhesion of titanium work onto them gradually reduces the pin height and increases the applied load.

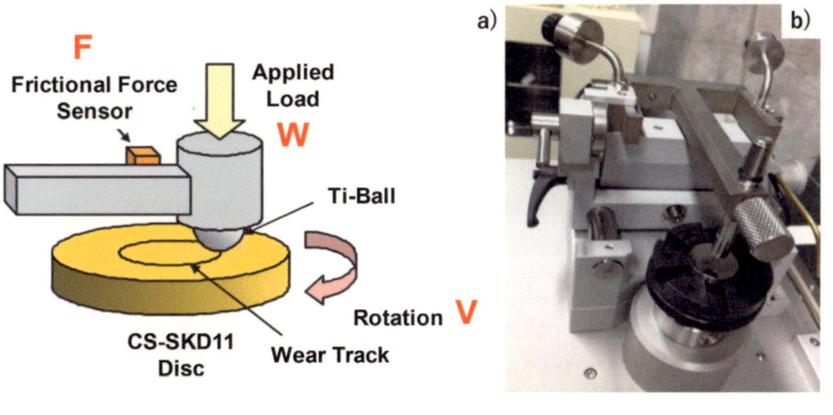

Figure 4.
Pure titanium ball to CS-SKD11 disc testing. a) Its schematic view of the rotational sliding test and b) overview of the whole setup.

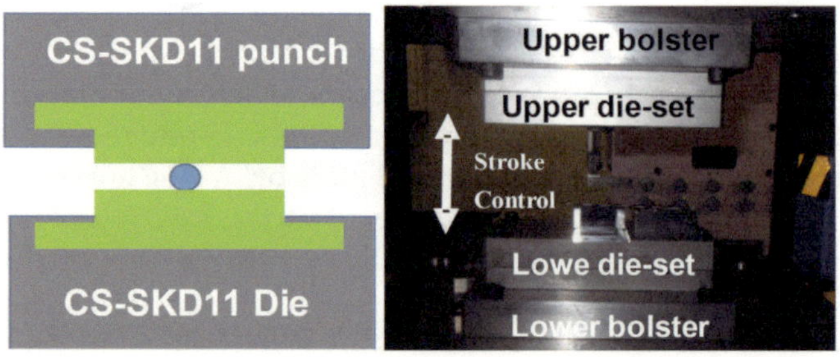

Figure 5.
Schematic view and overview of the upsetting process to uniaxially compress the titanium bar by CS-punch and die.

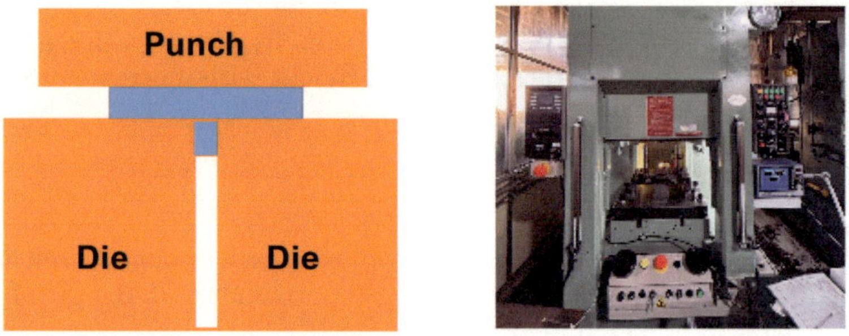

Figure 6.
Schematic view and overview of the pin-formation process to build up the pins into the titanium alloy frame.

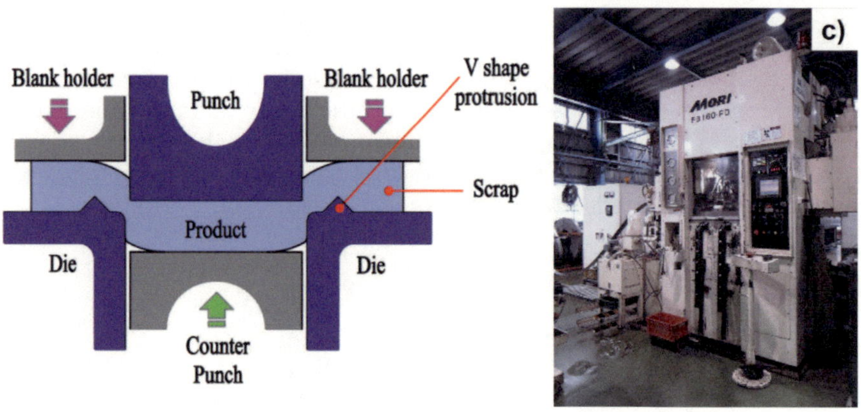

Figure 7.
Schematic view and overview of the fine blanking process to shear the thick titanium plate and to punch out the titanium product.

Figure 7 shows the fine blanking process to shear out the titanium plate to bulk products under constraints and various applied loads. During this shearing process, the fresh titanium work has a risk of galling to the punch and die surfaces. This adhesive wear significantly reduces the punch life and lowers the product quality.

3. Tribology of CS-SKD11 dies

Variation of the measured friction coefficient (μ) with the sliding (L) was depicted in **Figure 6** under the applied load (W) of 10 N and the sliding velocity (V) of 0.05 m/s. Except for the initial running distance, μ was preserved at 0.11 in average. μ and its deviation (Δμ) decreased from 0.15 to 0.10 and from 0.05 to 0.01, respectively. As discussed in [6], titanium debris stuck to TiN, TiCN and TiAlN coatings, resulting in higher friction coefficient than 1.0. This severe metal sticking was also observed on the wear track of Si-bearing DLC coating. No sticking was seen when using the nano-laminated DLC (n-DLC) coating; μ is nearly constant by 0.2. The surface textures and tribological conditions were varied to analyze their effect on the frictional transients in [30]. In every case, μ is more than 0.18. The stable and low-frictional state was preserved at CS-SKD11 disc against the pure titanium ball in dry.

SEM–EDX was utilized to analyze the wear-track surface of this CS-SKD 11 disc after sliding up to L = 500 m. As shown in **Figure 8**, no adhesion of pure titanium fragments was detected on this wear track except for a few TiO_2 debris particles at

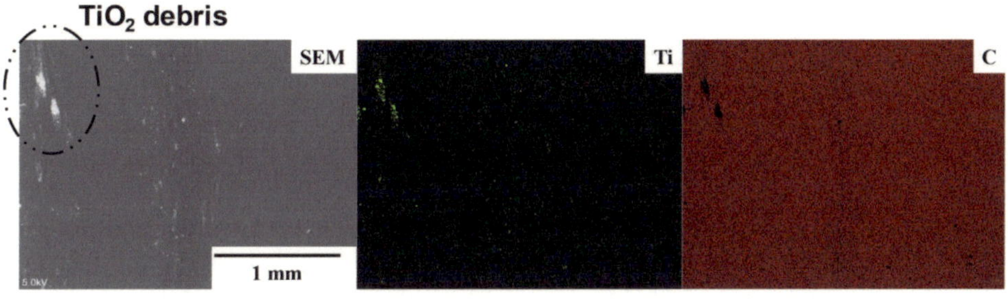

Figure 8.
SEM–EDX analysis on the wear track of CS-disc after sliding the pure titanium ball up to L = 500 m.

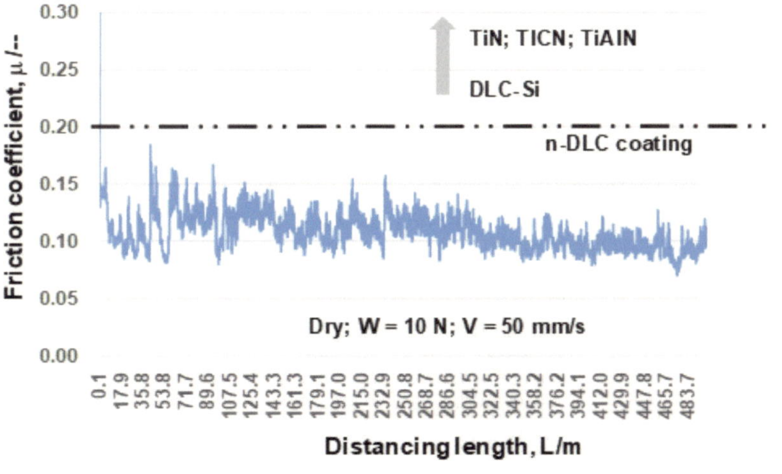

Figure 9.
Variation of the friction coefficient (μ) with increasing the sliding distance (L) up to L = 500 m under the applied load of 10 N and the sliding velocity of 0.05 m/s.

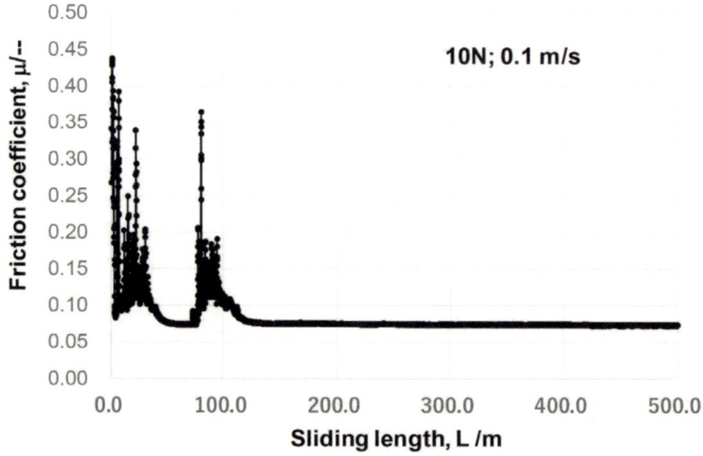

Figure 10.
Effect of the sliding velocity on the transients of friction coefficient with L. the applied load was 10 N and the sliding velocity was 0.1 m/s.

the vicinity of wear-track edges. Most of wear-track surface was covered by carbon film. This analysis proves that low-frictional behavior in **Figure 9** is attained by the in situ formation of free-carbon tribofilm onto the interface between CS-SKD11 disc and titanium ball. Let us consider the effect of sliding velocity on this low-frictional behavior. **Figure 10** shows the variation of friction coefficient with increasing L under W = 10 N and V = 0.1 m/s. In the initial running, spiky increase of friction coefficient was detected twice before L = 100 m. For L > 100 m, the stable friction advanced by μ = 0.07. This implies that a stable free-carbon tribofilm is formed after an incubation time when the insufficient coverage of free-carbon film at the hot spots induces unstable surface conditions with high friction coefficients.

4. Upsetting of titanium and titanium alloys by CS-dies

A pure titanium round bar with the diameter of 1.0 mm was upset in dry without any liquid lubricants by CS-punch and CS-die in the specified reduction of thickness (r). Before upsetting, the surfaces of CS-punch and CS-die were completely polished not to leave any surfactants. At first, this pure titanium bar was upset in a single shot by r = 10%, 20%, 30%, 50% and 70%, respectively. Without carbon supersaturation treatment, this reduction of thickness was significantly reduced down to 10 to 15%.

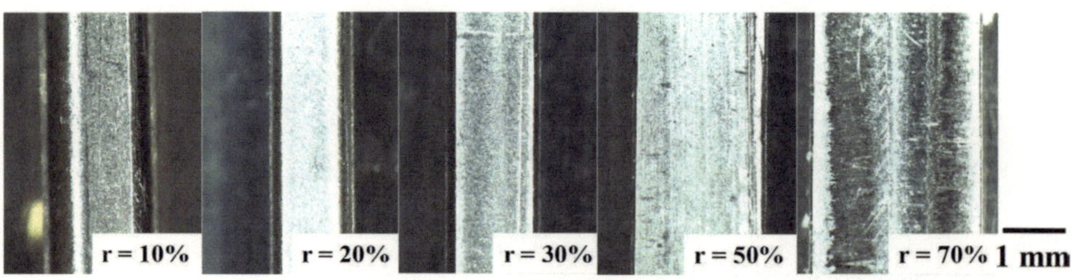

Figure 11.
Variation of the upset pure titanium round bar with increasing the reduction of thickness (r) up to r = 70%.

Aa shown in **Figure 11**, the round bar was upset to each r without adhesive wear on the contact interface.

The width of upset bar (W_o), the contact interface width (W_i) and the bulging displacement, B_g (=(W_o-W_i)/2) were chosen as a parameter to describe this upsetting behavior. After [31], B_g is employed as a parameter to measure the frictional coefficient during the uniaxial compression. When $B_g > 0.5$, the friction coefficient governs the plastic deformation of work and a round bar significantly bulges by itself. On the other hand, the friction coefficient reduces when $B_g < 0.5$ or $B_g \sim 0$. **Figure 12** shows the variation of W_o, W_i and B_g with r. W_o increases with r in quadratic, while W_i increases in linear with r. When r > 50%, W_i approaches W_o; at r = 70%, $B_g \sim 0.0$.

This monotonous approach of Wi to Wo implies that the friction coefficient monotonically reduces itself with r and goes down to nearly zero. Then, Bg monotonously decreases with r and becomes nearly zero at r = 70%. After the regression curve between Bg and μ, μ is estimated to be 0.05–0.1, much lower than the friction coefficient with μ > 0.2 in the literature [6, 30]. This low-frictional behavior by CS-punch and die against the titanium bar is reconsidered by using the finite element analysis (FEA) [24]. In this inverse analysis, the friction coefficient is parametrically varied so that the measured micro-hardness map corresponds to the plastic strain calculated by FEA at the same reduction of thickness. **Figure 13** compares the measured micro-hardness map to the calculated plastic strain at r = 50%, assuming that the friction coefficient is 0.05; that is, μ = 0.05 at r = 50%. Let us analyze the work-die interface using the SEM and EDX, where low friction is preserved even with increasing r. The contact interface area expands from the center line toward the upper and lower directions with increasing r. Then, without CS, the fresh titanium work could be easily in direct contact with the die surface to onset the galling. In this contact area, fine white stripes are seen along this area expansion in **Figure 14a**. EDX was utilized to identify the chemical component of these strips at the central region A in the contact interface. As compared in **Figure 14b**, no correlation is recognized between these stripes and

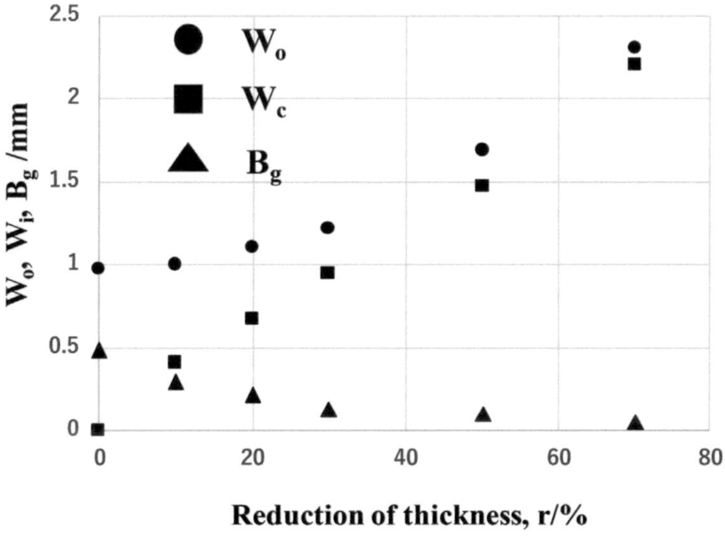

Figure 12.
Variation of the bar width (W_o), the contact interface width (W_c) and the bulging displacement (B_g) with increasing r.

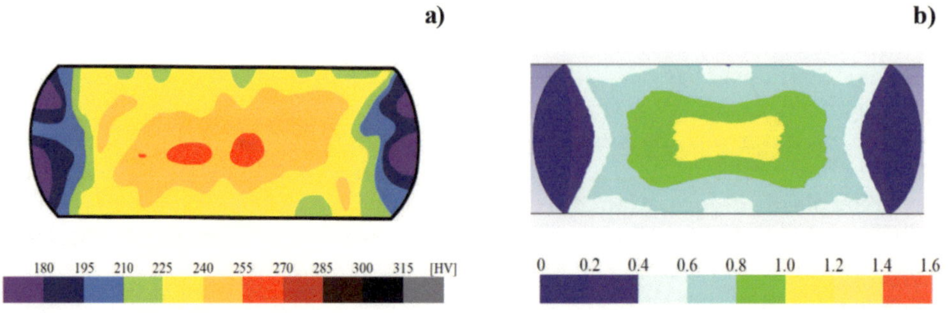

Figure 13.
Comparison of the experimentally measured micro-hardness map with the calculated plastic strain distribution by FEA. a) Measured micro-hardness map and b) calculated plastic strain.

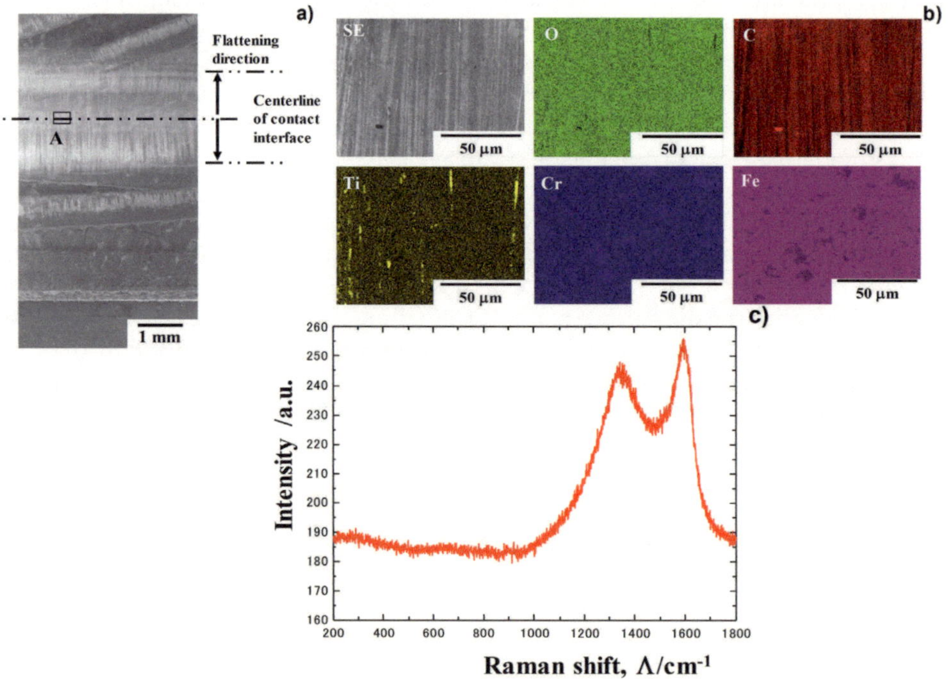

Figure 14.
Microstructure analysis on the contact interface between the CS-punch and the titanium bar after continuously upsetting the pure titanium bar 10 times up to r = 70%. a) SEM image on the contact interface, b) element mapping by EDX at the region a in (a, and b) Raman spectroscopy at the region a in (a).

constituent elements of SKD11, iron and chromium. A very few stripes correspond to the titanium and oxygen maps; the TiO_2 debris particle sparsely adhered to the CS-punch surface. Almost all of white stripes in SEM images just correspond to carbon mapping. Since no stripes were detected before upsetting, these carbon stripes must be in situ formed on the interface. Raman spectroscopy was utilized to describe the binding state of these carbon white stripes. Since no significant peaks were found in the lower wave numbers in **Figure 14c**, the carbon was not present as a bound carbon in the carbides. Fine D-peak at Λ = 1320 cm^{-1} and G-peak at Λ = 1600 cm^{-1} prove that these stripes consist of free carbons, isolated from the CS-punch during upsetting.

As already discussed in [9, 32], a free carbon has no possibility to isolate from amorphous carbon layer or DLC coating at the absence of polymers and liquid lubricants. This implies that a free carbon directly isolated from CS-SKD11 die drives the low friction and wear condition at the hot spot in the contact interface to titanium works.

Let us compare the upsetting behavior between pure titanium and β-phase titanium alloy bars using the same CS-punch and CS-die. As suggested in [33], the pure titanium has hcp (Hexagonal close-packed) structure, while the β-phase titanium alloy has bcc (body-centered cubic) structure. Due to a fewer number of slipping planes, hcp-structured material is thought to have more difficulty in plastic straining than bcc-structured one. This difference is reflected in the upsetting behavior under nearly the same low-frictional interface to CS-punch and CS-die.

The micro-hardness mapping is employed to compare the plastic straining behavior between two bars under the same upsetting conditions as shown in **Figure 15**. Even when r = 10%, the β-phase titanium alloy bar has higher micro-hardness profile than the pure titanium bar. This reveals that the uniform plastic straining advances in the β-phase alloy more easily than the pure titanium. In both bars, work hardening is suppressed only at the center of bars in common since the CS-punch and CS-die are utilized in both cases. The hardness gradient, that is, the plastic strain gradient is much reduced even at higher reduction of thickness when using the β-phase alloy. This implies that uniform plastic straining with lower work hardening advances in upsetting

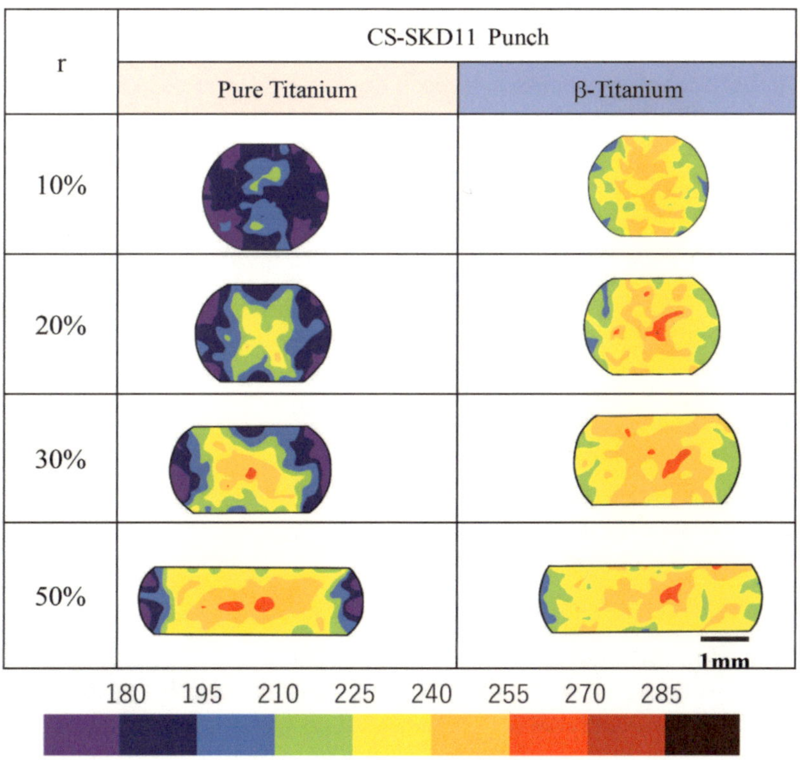

Figure 15.
Comparison of the micro-hardness mapping between the pure titanium and -phase titanium alloy bars, using the same CS-punch and CS-die.

the β-phase titanium alloy. Hence, the upsetting of β-phase titanium alloy is free from the shear localization and ductile fracture even at the high reduction of thickness.

The upsetting process is widely utilized in metal forming to imprint the die cavity shape to an upset work under free forging situation and to shape the pins onto the forged titanium products. Let us apply this CS-treatment to free forging punch and die. In practical upsetting process, the die is often shaped to form the initial round titanium bar to the deformed cross-sectional bar. In the conventional process without carbon supersaturation treatment, the shaping stage must be divided into several steps including the polishing and surface treatment with intermission. The CS-treatment is expected to save those tedious steps and to make near-net shaping from the round bar to the product in a single stroke. In the following experiment, the initial round bar with the diameter of 3.0 mm is free forged to the triangle cross-sectional bar within a single stroke.

The CS-punch and CS-die in **Figure 16** were prepared for upsetting the round pure titanium bar to a triangular cross-sectional bar. **Figure 17** describes the variation of bar shape with increasing r up to 45%. Through the swelling of work into the die cavity for $r < 10\%$, the titanium work gradually fills into the whole die cavity with increasing r for $r > 20\%$. Let us measure the load-stroke relationship at each reduction of thickness. After the initial swelling for $r < 10\%$, the load gradually increases with contact interface area extension with r, as shown in **Figure 18**. This stable load-stroke relationship proves the monotonous filling of the swelled titanium bar into a die cavity to form the triangular cross-sectional bar.

Let us consider the role of CS-die on the filling process during shaping of the round β-phase titanium alloy bar to the triangular cross-sectional bar. HT (heat treatment) SKD11 die was used as a reference to analyze this role. The same CS-punch was used in both shaping processes. In both cases, no difference in the plastic strain distribution is noticed from the initial swelling step to the early stage of filling step into a die cavity in **Figure 19**. Toward the final stage of filling step, the uniform straining behavior is lost when using the HT-die, so that work hardening severely occurs at the vicinity of CS-punch surface and HT-die. On the other hand, the β-phase titanium work becomes free from this local work hardening and deforms with sufficient uniformity in straining when using the CS-die.

Finally, let us apply this pair of CS-punch and CS-die to continuous shaping of round β-titanium round bar to the triangular bar by shifting the CS-punch in every 20 mm.

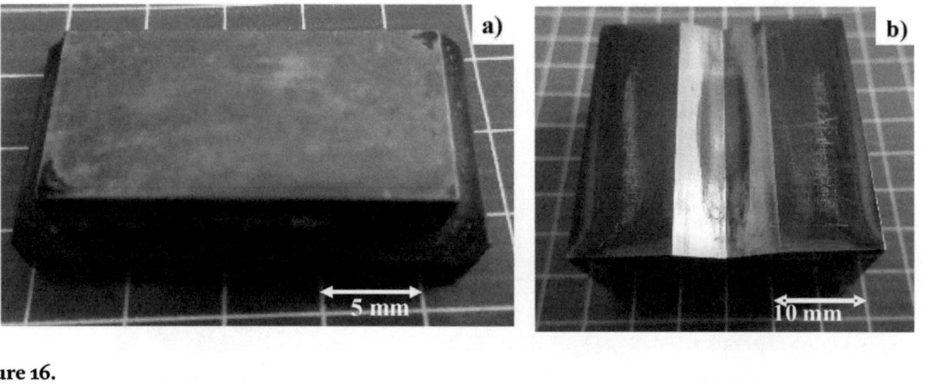

Figure 16.
CS-punch and CS-die for upsetting the round bar with the diameter of 3.0 mm to a triangular cross-sectional bar. a) CS-punch and b) CS-die.

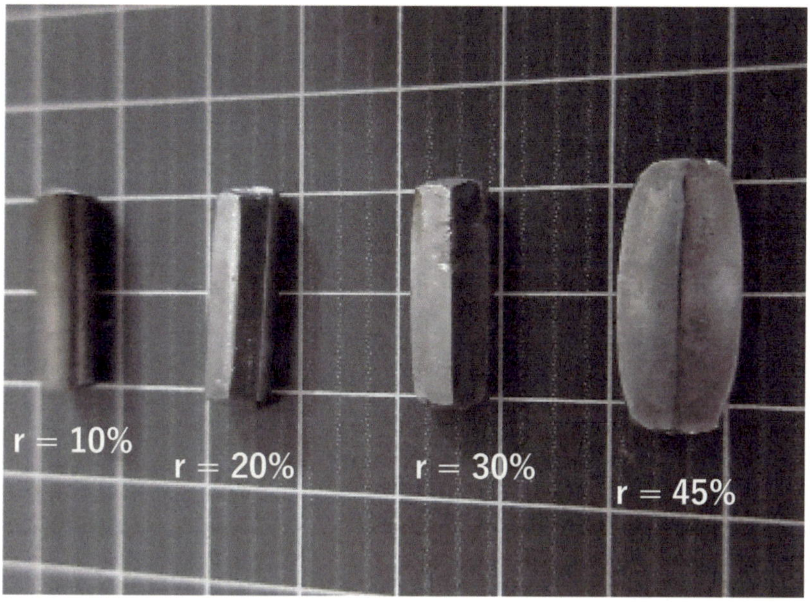

Figure 17.
Variation of the work shape with increasing the reduction of thickness (r) up to r = 45%.

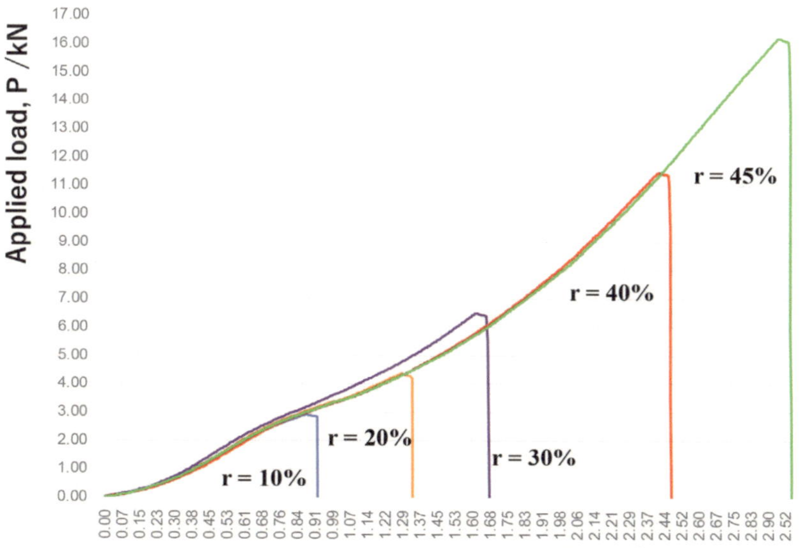

Figure 18.
Transients of the load-stroke relations with increasing r up to r = 45%.

Figure 20 compares the initial round bar to the diameter of 3.0 mm with the continuously upset bar in four times. Within a single shot in each upsetting step, the initial round bar is successfully transformed to a long triangular bar. This demonstrates that a round β-titanium bar is used as a feedstock to deform any cross-sectional stringer and frame by the incremental shaping in tandem procedure.

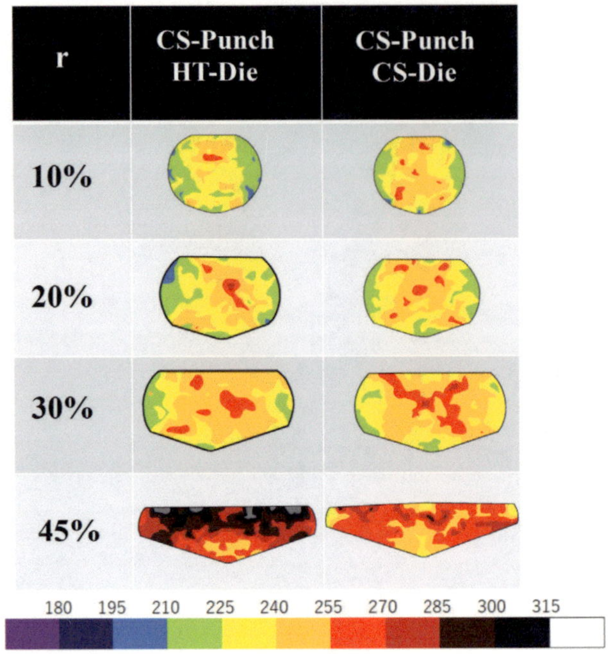

Figure 19.
Transients of the micro-hardness mapping in cross-section of β-phase alloy bars during the shaping process with and without the use of CS-die.

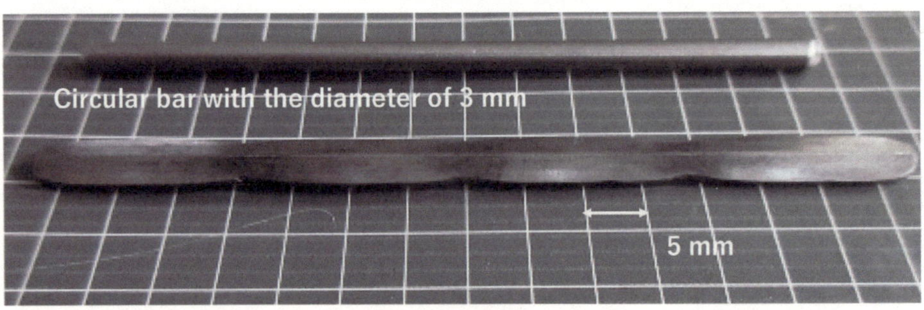

Figure 20.
An incremental shaping with r = 45% to continuously deform the round bar to a triangular cross-section bar by shifting the CS-punch in each 20 mm.

5. Titanium alloy pin forming by CS-die

The titanium eye-glass frame is made of various parts, which are forged and shaped all together in dry without lubrication [34]. A pin is a typical element to fix the interior component to titanium frame for high qualification of product. Due to the galling of titanium alloy to the dies in dry, this pin forming must be scheduled in multi-steps. Each unit step consists of the heat treatment and pickling, the anodizing and demold-coating and the upsetting. After this upsetting, the additional sub-steps are often necessary for trimming and barreling the burs. Toward high qualification of products, the first upset pin height has a significant influence on the final pin height with sufficient aspect ratio after multi-step operation.

In this titanium pin forming, the first upset pin height is employed as a parameter to describe the superiority of workability to the CS-dies. Tall pin height is preserved all through the pin-forming process. In particular, the intermediate steps and sub-steps are significantly saved by using the CS-die.

A type ACD56 tool steel with the hardness of HRC57 was prepared as a die for pin forming. A pure titanium bar with the diameter of 5 mm was used as a feedstock. A knuckle-joint stamper (AIDA160) was utilized for upsetting with the applied load of 1200 kN.

Figure 21 compares the first upset pin height by using the normal ACD56 die and CS-ACD56 die. The first upset pin height by the normal die is limited by 1.22 mm, less than the targeting height of 1.35 mm in schedule. On the other hand, the first upset height by CS-die reaches 1.68 mm, much taller than the targeting height. This difference proves that low friction and low adhesive wear are attained by using the CS-die just in the similar manner to the free forging process with the use of CS-punch and CS-die.

When using the normal die, its surface was adhered by the fresh titanium fragments. This galling increases the friction on the interface between the die and titanium work. If the applied load were kept constant, the upset height could decrease with increasing the number of strokes.

As expected in the above, the first upset ping height (h) gradually reduced with N when using the normal die; e.g., h = 1.22 mm at N = 1, h = 1.0 mm at N = 20 and h = 0.94 mm at N = 35. In addition to this reduction of pin height by galling, the deviation of h is also enhanced with increasing N, as depicted in **Figure 22**. On the other hand, h is nearly constant by h = 1.67 mm with less deviations even with increasing N. This high workability, with indifference to N, demonstrates that in situ formed free-carbon tribofilm significantly reduces the friction and adhesive wear on the contact interface.

The first upset pin height is preserved to be higher than the target level, so that the successive steps and sub-steps can be saved as pointed in **Figure 23**. In case that the target height is 3 mm, the first and second steps are enough to attain the high qualified products. In particular, the total processing cost might be significantly saved since many sub-steps for heat and surface treatments and surface polishing become unnecessary.

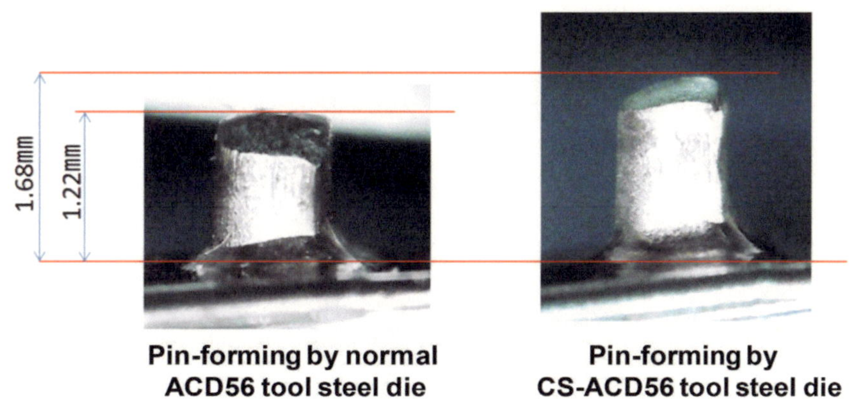

Figure 21.
Comparison of the first upset pin height by the normal and CS-ACD56 dies.

In general, the forming procedure of titanium and titanium alloy bars to the final frame product consists of time-, energy-, labor- and cost-consuming processes. This reduction of steps and sub-steps in the pin-forming process enables a shortened processing time, reduced energy consumption and savings in labor cost

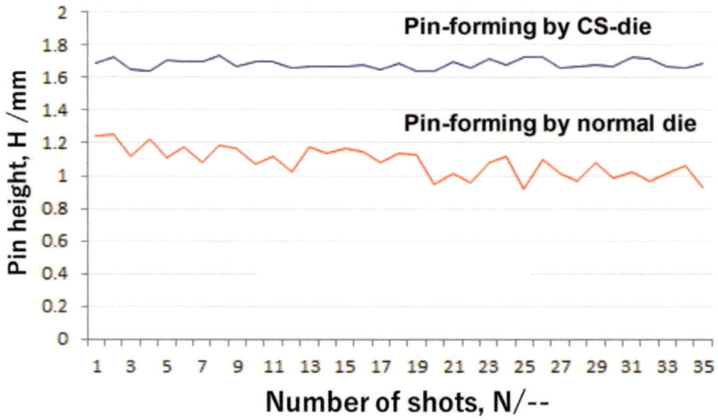

Figure 22.
Variation of the first upset pin height with increasing the number of strokes, using the normal and CS-dies.

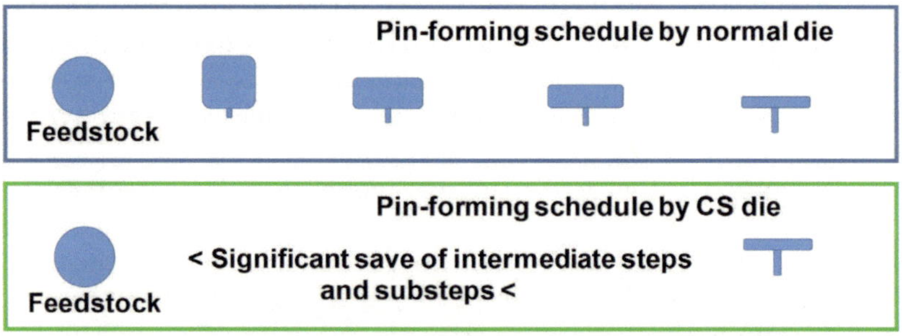

Figure 23.
Saving the intermediate steps in the pin forming when using the CS-dies.

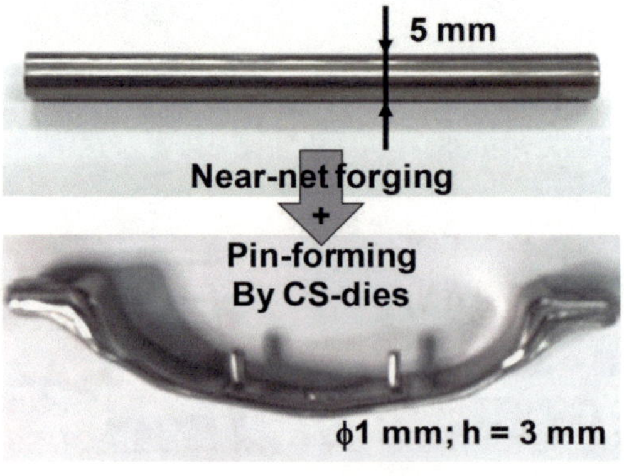

Figure 24.
New pin forming with near-net forging process in the procedure from the feedstock to the glass-frame preform.

in other parts of the forming schedule. In addition, the distance and gap between the designer's product and the formed eye-grass frame can be much reduced, so that value of eye-glass frame products is much improved by manufacturing the designer's model.

Figure 24 shows the revised pin-forming process with the near-net forging from the feedstock. Under the significant surface extension of work materials, low friction and wear state are preserved by using the CS-die.

6. Fine blanking of pure titanium plates by CS-dies

In the fine blanking, the metal, the polymer and their composite plates were sheared under the mechanical constraint to punch out the near-net-shaped parts with nearly full burnished surface [35]. As seen in **Figure 7**, various loads are also applied to fix the work plates between the dies and the work. Then, the ductile and active metals and alloys, such as the oxygen-free copper, the pure titanium and the stainless steels, are easy to adhere onto the punch and die surfaces [36].

Figure 25 shows a typical adhesion of debris fragments when fine blanking the pure titanium plate with the thickness of 2.0 mm even under lubrication. Since this galling occurs in a single-shot blanking, an innovative tooling to suppress the direct contact of fresh titanium work to dies is indispensable to continue the blanking process in the continuous production of the titanium parts. The CS-YXR7 (matrix type tool steel) punch and die are utilized to fabricate the pure titanium gears [37, 38]. The fine blanking stamper (FB-160FB; Mori Iron Works, Co., Ltd., Fukuoka, Japan) was employed to make continuous fine blanking experiments. The CS-YXR7 punch was machined and finished to have 12 teeth module-1.0 involute spur gear geometry after CAD-data with the pitch circle diameter of 12.00 mm, the pressure angle of 20°, the tip circle diameter of 14.00 mm, the tip radius of 0.3 mm, the root circle diameter of 10.00 mm and the root radius of 0.4 mm, respectively. These punch edges were accurately finished to have double chamfers for suppressing their abrasive wearing.

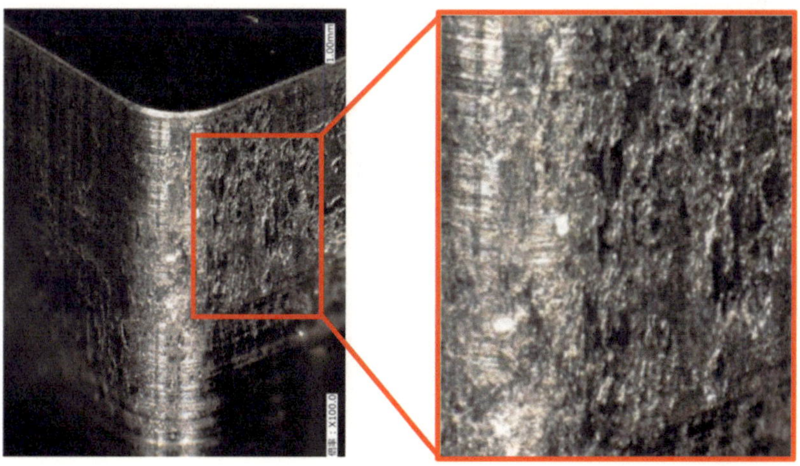

Figure 25.
Galling of pure titanium debris fragments onto the sheared surface of YXR7 punch with the square head.

Figure 26 depicts the sheared-out titanium gear blanks. Smooth gear surfaces are attained by using the CS-punch and die. This proves that severe adhesion of fresh titanium debris did not occur in this fine blanking process with the use of CS-YXR7 punch. How about the continuous fine blanking of pure titanium work? **Figure 27** demonstrates that the series of titanium gears is yielded without severe damages to titanium blanks by galling. To be noticed, the gear-tooth surfaces of sheared blank at N = 1 have a fracture surface in partial. This partially fractured surface area fraction decreases with increasing N; the blanked gear-tooth surfaces are fully burnished at N = 50.

Once the galling occurred on the punch to work interface at the initial or intermediate stages in this continuous fine blanking, the fractured surface area fraction could be gradually or steeply enhanced with increasing N. **Figure 27** also reveals that the CS-punch to titanium work interface must be solid lubricated to reduce the fractured surface fraction by in situ formation of free-carbon tribofilm [39].

Let us investigate the surface conditions of tooth surface at the CS-YXR7 punch. **Figure 28** compares the tooth edge profile of CS-YXR7 punch at the initial stage and after fine blanking up to 50 strokes. Very thin fragments with the thickness of clearance are seen on the tooth surfaces of CS-punch after shearing at N = 50. This punch was softly polished by the polymer brushes. Those thin fragments are cleaned and

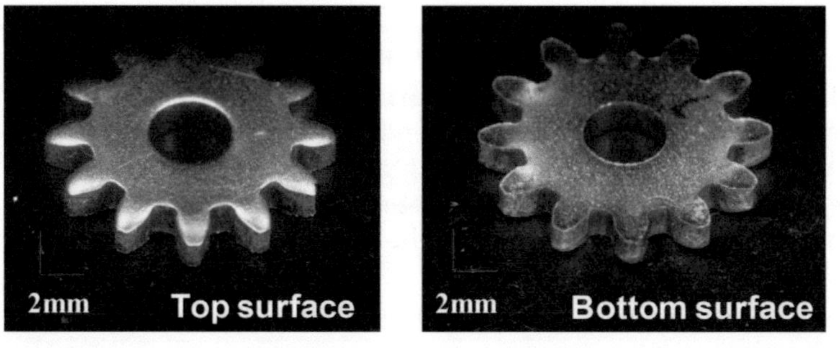

Figure 26.
The sheared-out pure titanium gear blank. Droops and burrs were slightly detected on the top and bottom blank surfaces, respectively.

Figure 27.
Continuous fine blanking of pure titanium gears up to N = 50. The gear side surface condition was improved with increasing N.

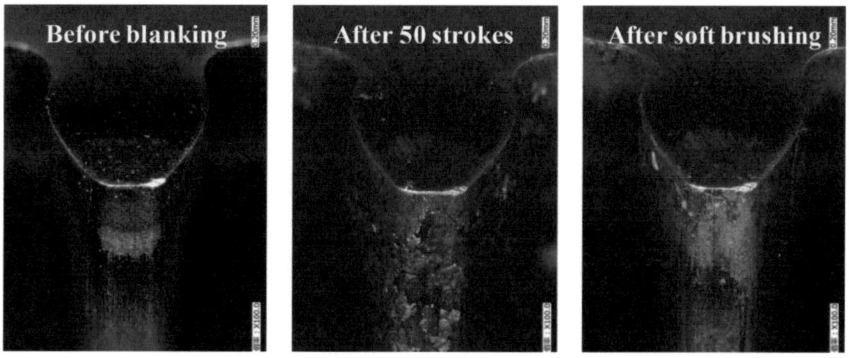

Figure 28.
Comparison of a gear-tooth profile before fine blanking, after fine blanking up to N = 50, and after softly brushing the used punch surface at N = 50.

polished away, so that the surface condition of CS-YXR7 after brushing comes back to the initial condition, as compared in **Figure 28**. This proves that no galling takes place to induce the damages to the punch surfaces.

Surface roughing process on the gear blanks is analyzed with increasing the number of strokes. **Figure 29** depicts the variation of average surface roughness (Ra) with increasing N. Two positions were selected on the gear side surfaces at the 0 and 6 o'clock angles. Ra steeply increases from 1 to 2 μm up to N = 2. For N > 2, this roughness gradually decreases with N; e.g., Ra decreases from 2 μm at N = 2 to 0.6 μm at N = 50 at 0 o'clock angle. This initial steep increase corresponds to the occurrence of the fractured surface on the gear side surface as shown in **Figure 27**. The monotonous decrease of Ra with N is also in agreement with the gradual decrease of fracture surface area fraction with N. That is, the free-carbon tribofilm does not cover the whole contact interface to titanium work at N = 2, so that the fresh titanium fragments are taken away from work and deposited onto the CS-YXR7. This surface state change

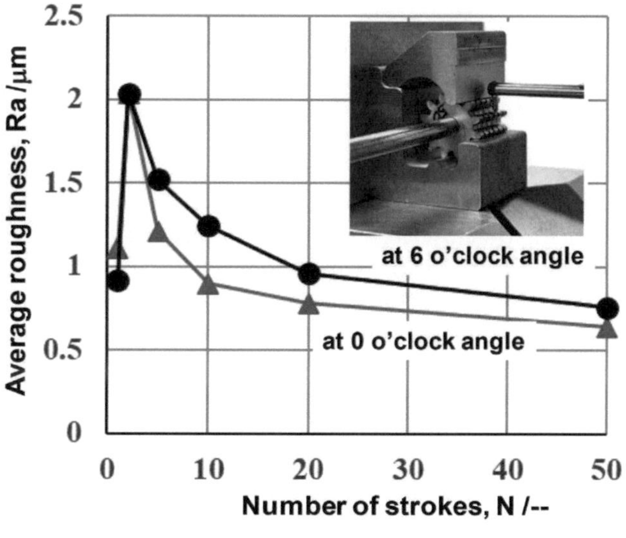

Figure 29.
Variation of the sheared gear-blank surface roughness at two positions with increasing the number of strokes.

reflects on the surface roughing of blanks. With increasing the number of strokes, the free-carbon films agglomerate and cover the hot spots on the contact interface to be free from galling. Then, the burnished surface area fraction increases and the surface roughing is gradually suppressed to be less than the initial roughness level.

7. Discussion

The galling or the adhesive wear becomes one of the most difficult issues to be overcome in the metal forming design. As reviewed in [40], four engineering items affect this problem in a complex way; i.e., the plastic strain, the surface condition, the flash temperature and the work material. This galling is classified into two categories: the mechanical galling and the mechano-chemical galling.

In the former issue, the friction and wear of tools and dies can be reduced by the surface treatment like ceramic coating or DLC coating [41], by the reduction of surface roughness [42], or by optimal selection of lubricants [43]. However, the latter issue is difficult or nearly impossible to be solved only by those methods. A typical mechano-chemical galling is experienced in forging, shaping and shearing the titanium and titanium alloy works. As reported in [43], the anodizing method played a role to make drawing and bending the titanium sheets without galling, even in MQL (Minimum Quantity Lubrication), since the contact area is limited to be narrow under a little application of plastic strains.

Severe mechano-chemical galling occurs in the forming of titanium alloy eye-glass flames, in the near-net forging of titanium feedstock to preforms or in the fine blanking of titanium plates to mechanical parts. The true contact interface between work materials and dies must be in situ lubricated in solid all through these processes. The massively carbon supersaturated (CS) tools and dies provide a non-traditional way to significantly reduce this type of galling in practical forming processes of titanium and titanium alloys.

In situ formation of free-carbon tribofilm onto the hot spots in the contact interface is common to every tribotesting and metal forming with the use of CS-disc, CS-punch and CS-die, respectively. The massively supersaturated carbon in the nano-structured layer is driven by the stress gradient to make jumping diffusion from one octahedral vacancy sites to the other through the nano-cluster boundaries, to isolate itself from the layer of CS-tool and to agglomerate on the hot spots [44]. Just in the similar manner in the nucleation and growth theory in materials science [45], an incubation time is necessary for the isolated free carbon to nucleate on the hot spots, to agglomerate itself to a solid dot and to form the free-carbon tribofilm on the contact interface. During this intermission, the titanium and titanium alloy work has a risk to adhere onto the CS-tool surface in partial.

In case of the tribotesting, the sticky increase of friction coefficient was observed at the initial stage in sliding movement of titanium ball onto CS-disc as seen in **Figures 9** and **10**. In the upsetting and free forging processes, the local adhesion of debris particles was seen on the hot spots as shown in **Figure 12**. At the beginning of the pin formation, a local adhesion of titanium alloys was detected in practical operations. The titanium adhesion was detected at the vicinity of CS-punch edge when fine blanking the pure titanium plates. This rate-effect on the local occurrence of adhesion could be saved by initially applying the stress gradients to the CS-tools before actual metal forming. During this learning step, the in situ solid lubrication process is activated in the CS-tools. Mechanical tapping [46] provides a way to form a thin free-carbon tribofilm onto the specified surface regions of CS-tool. Then, the tribofilm could work from the initial contact state to titanium and titanium alloy work materials.

The titanium alloy design stood on the optimum selection of chemical components to modify the original hcp-structured crystallography to bcc- or other structured system by phase transformation [47–49]. Although those studies were based on the uniaxial tensile or compressive experiments, the workability of designed titanium alloys was not evaluated properly under the multi-axial plastic straining. As studied in [50], the friction on the tool-work interface significantly affected the plastic strain localization and the ductile fracture in compression. The plastic strain gradient and local work hardening also have much influence on the workability of titanium alloys. **Figures 15** and **19** demonstrate that low friction and low work hardening are sustained even at the high reduction of thickness, using the CS-punch and CS-die. Under this metal forming condition, the homogeneous plastic straining advances with active slipping systems of β-phase alloys. The superiority to β-phase alloy in plastic straining in the alloy design comes true only when using these CS-punches and CS-dies.

Let us consider the extension of this CS-die technology to the grain-size refinement and the forging of other work materials than titanium.

The grain size reduction with high crystallographic misorientations provides the simplest way to improve the strength of titanium and titanium alloys. Among various procedures for grain size refinement, the repeated forging process with the use of CS-dies becomes a suitable procedure to refine the original coarse grain size before near-net shaping.

The stress gradient to drive the isolation of carbon solute from nanostructure is induced by traction to the CS-tool surface during the elasto-plastic deformation of work materials. Then, the flow stress of work materials has influence on this applied traction to contact interface. When using the soft metals such as tin, lead or pure aluminum, this in situ solid lubrication has a risk not to be working, since the sufficient amount of free-carbon dots is not supplied from the CS-tools. In each material system, the initial learning process is needed to activate the in situ solid lubrication mechanism before actual forming. Almost all the stainless steels are in situ lubricated in solid without the initial learning step. When using AISI304 plates in fine blanking, the AISI304 gears are punched out without adhesive wear just as seen in **Figures 26** and **27** [51]; much better burnished gear surface is attained by CS-punch and CS-die.

The light elements, such as boron (B), carbon (C), nitrogen (N) and oxygen (O), are possible to make massively supersaturation into various material systems, to significantly modify the initial microstructure of punch and die materials to nanostructures and to form a tribo-chemically synthesized surface layer. After the theoretical study on the supersaturation process [52], the nanostructuring with cluster formation is common to these massively interstitial solute supersaturation. The chemical activity is different in each species of interstitial solute. In the present study, the nanostructure of free-carbon film is synthesized from the diffusing carbon solutes through the nano-cluster boundaries. This chemical reaction on the contact interface of supersaturated dies to work materials is controlled by the chemical affinity between the interstitial solutes and the constituent elements of work materials.

8. Conclusion

The metal forming of titanium and titanium alloy parts, members and components for industries and medical markets has suffered from severe galling to dies by their debris fragments and particles. In particular, long forming steps with high labor cost and energy consumption hinder the way to advance the manufacturing process in their precise forging, net-shaping and fine blanking. The carbon supersaturation with high carbon solute content prevents these metal forming punches and dies from severe adhesions, so that the titanium and titanium alloy products are formed with high dimensional accuracy and with high surface finish.

The incremental dry forming of eye-glass titanium alloy frames required long series of steps including the additional sub-steps such as heat treatment, surface polishing and deburring. Five steps including these sub-steps are reduced to two steps without any sub-steps by using the CS-punch and die. This reduction of incremental forming steps and sub-steps significantly lowers the labor cost and energy consumption. The continuous mass production of titanium gears by fine blanking is free from intermission by adhesion of titanium work to punch and die. The lubricating oil consumption is much reduced to MQL level. Fully burnished gear surfaces are maintained through the whole production schedule.

In common to the above industrial applications, low friction on the contact interface results in the reduction of applied load. Low wearing state preserved through manufacturing proves high quality of product. Low work hardening assures the flexibility in near-net shaping and provides the way to complex shaping by precise forging. This advanced manufacturing by this carbon supersaturation technology must be essential in the sustainable society with circulation economy.

Except for the initial learning stage, the metal forming performance as well as the product quality are improved with increasing the number of strokes by using the CS-tools. This self-lubrication by the free-carbon tribofilm from CS-tools assures the longer tool life without change in the product quality. The incubation time for initial learning step is reduced by the mechanical tapping or by the learning trials. Since the isolation and diffusion of free-carbon solute from CS-dies to hot spots on the die-work interface are driven by the stress gradient, the formation of free-carbon tribofilm and its coverage onto hot spots is strongly dependent on the flow stress of work materials. BOD-testing and upsetting experiments are necessary to describe how effectively this in situ solid lubrication plays a role in each work-material system.

Massive carbon supersaturation via the low temperature plasma carburizing is further studied to deepen the CS-layer thickness in various die materials, to control the CS-layer distribution in depth and the hardness depth profile and to make plasma-printing with assistance of carbon solutes.

Acknowledgements

The authors would like to express their gratitude to S-I. Kurozumi (Nano-Coat Film, llc.), and S. Ishiguro (Graduate School of Engineering, Toyama University; now, PANASONIC, Co., Ltd.) for their help in experiments.

Conflict of interest

The authors declare no conflict of interest.

Author details

Tatsuhiko Aizawa[1*], Kenji Fuchiwaki[2], Takeshi Kihara[3] and Tomomi Shiratori[4]

1 Surface Engineering Design Laboratory, Shibaura Institute of Technology, Tokyo, Japan

2 Hatano Precision, Kanagawa, Japan

3 Masunaga Optical Mfg. Co., Ltd., Fukui, Japan

4 Faculty of Engineering, University of Toyama, Toyama, Japan

*Address all correspondence to: taizawa@sic.shibaura-it.ac.jp

References

[1] Beal JD, Boyer R, Sanders D. Forming of titanium and titanium alloys. In: ASM Hand-Book 14B: Metal Working, Sheet Forming. Materials Park, OH, USA; 2006. pp. 656-669

[2] Hutchings I, Shipway P. Tribology: Friction and Wear of Engineering Materials. 2nd ed. New York, USA: Elsevier; 2017

[3] Kihara T. Visualization of deforming process of titanium and titanium alloy using high speed camera. In: Proc. 2019-JSTP Conference. Tokyo, Japan: JSTP; 2019. pp. 41-42

[4] Akbarzadeh M, Zandrahimi M, Moradpur-Tari E. Molybdenum disulfide coating on AISI316 stainless steel by thermo-diffusion method. Archives of Metallurgy and Materials. 2017;**62**:1741-1748

[5] Kataoka S, Murakawa M, Aizawa T, Ike H. Tribology of dry deep-drawing of various metal sheets with use of ceramic tools. Surface and Coating Technology. 2004;**178**:582-590

[6] Dohda K, Aizawa T. Tribo-characterization of silicon doped and nano-structured DLC coatings by metal forming simulators. Manufacturing Letters. 2014;**2**:82-85

[7] Dohda K, Yamamoto M, Hu C, Dubar L, Ehmann KF. Galling phenomena in metal forming. Friction. 2021;**9**:665-685

[8] Ahmed A, Li C. Lubrication and surface engineering. In: Principles of Engineering Tribology. Ch. 7 ed. Cambridge, MA, USA: Academic Press; 2023. pp. 295-343

[9] Aizawa T, Akhadejdamrong T, Mitsuo A. Self-lubrication of nitride ceramic coating by the chlorine ion implantation. Surface and Coating Technology. 2004;**178**:573-581

[10] Mitsuo A, Uchida S, Yamamoto S, Aizawa T. Improvement of cutting performance for carbide tools via chlorine ion implantation. Surface and Coating Technology. 2004;**189**:630-635

[11] Sumitomo T, Aizawa T, Yamamoto S. In-situ formation of self-lubricating tribofilm for dry machinability. Surface and Coating Technology. 2005;**200**:1797-1803

[12] Bachmann S, Schulze M, Krell L, Merz R, Wahl M, Stark RW. Ultra-low friction on tetrahedral amorphous diamond-like carbon (ta-C) lubricated with ethylene glycol. Lubricants. 2018;**6**(3):59-68

[13] Aizawa T, Itoh K-I, Fukuda T. SiC-coated SiC die for galling-free forging of pure titanium. Materials Transactions. 2020;**61**(2):282-288

[14] Aizawa T, Yoshino T, Suzuki Y, Shiratori T. Anti-galling cold dry forging of pure titanium by plasma carburized AISI420J2 dies. Journal of Applied Sciences. 2021;**11**(595):1-12

[15] Aizawa T, Yoshino T, Suzuki Y, Shiratori T. Free-forging of pure titanium with high reduction of thickness by plasma carburized SKD11 dies. Journal of Materials. 2021;**14**(2536):1-12

[16] Aizawa T, Yoshino T, Fukuda T, Shiratori T. Dry cold forging of pure titanium wire to thin plate with use of β-SiC coating dies. Journal of Materials. 2020;**13**(3780):1-11

[17] Aizawa T, Ito K-I, Fukuda T. Galling-free micro-forging of titanium wire with high reduction in thickness by β-SiC dies. Forming the future. The Minerals, Metals & Materials Society. Pittsburgh, PA, USA; 2021;**1**:1065-1075

[18] Aizawa T, Yoshino T, Shiratori T, Dohda K. Anti-galling β-SiC coating dies for fine cold forging of titanium. Journal of Physics: Conference Series. 2021;**1777**:012043

[19] Aizawa T. Micro-manufacturing by controlled plasma technologies. Automotive Tech. 2017;**72**(6):35-41

[20] Aizawa T, Funazuka T, Shiratori T. In situ lubrication in forging of pure titanium using carbon supersaturated die materials. Journal of Nanomaterials. 2024 (in press)

[21] Aizawa T. Low temperature plasm nitriding of austenitic stainless steels. In: Stainless Steels and Alloys. Ch. 3 ed. London, UK, London: Intech Open, UK; 2019. pp. 31-50

[22] Aizawa T, Shiratori T, Yoshino T, Suzuki Y, Komatsu T. Nitrogen supersaturation of AISI316 base stainless steels at 673 K and 623 K for hardening and microstructure control. In: Stainless Steels. Ch. 4 ed. London, UK, London: Intech Open, UK; 2022. pp. 61-84

[23] Aizawa T. Micro–/meso-structure control of multi-hostmetal alloys by massive nitrogen supersaturation. Journal of Materials. 2024 (in press)

[24] Ishiguro S, Aizawa T, Funazuka T, Shiratori T. Green forging of titanium and titanium alloys by using the carbon supersaturated SKD11 dies. Journal of Applied Mechanics. 2022;**3**:724-739

[25] Aizawa T, Ishiguro S, Shiratori T, Yoshino T. Near-net forging of titanium and titanium alloys by the plasma carburized SKD11 dies. Key Engineering Materials. 2022;**926**:1143-1150

[26] Aizawa T, Funazuka T, Shiratori T. Near-net forging of titanium and titanium alloys with low friction and low work hardening by using carbon-supersaturated SKD11 dies. Journal of Lubricants. 2022;**120**(203):1-15

[27] Suenaga R, Yunata E, Aizawa T. Quantitative plasma diagnosis of high-density RF-DC plasmas for surface processing. In: Proc. 7th SEATUC Conference (March 7th, 2013; Bandon, Indonesia). Yogyajarta, Indonesia: Gajamada University Press; 2013. pp. 63-68

[28] Rosadi I, Djoko D, Sakti SP, Yunata EE, Aizawa T. Plasma diagnosis on the mixture gas plasma state for nitriding process. In: Proc. 12th SEATUC Symposium, 2018, March 3rd. Yogyajarta, Indonesia: Gajamada University Press; pp. 131-136

[29] Aizawa T, Rasadi I, Yunata EE. High density RF-DC plasma nitriding under optimized conditions by plasma diagnosis. Applied Sciences. 2022;**12**(3716):1-12

[30] Ding Q, Wang L, Hu L. Tribology optimization by laser surface texturing: From bulk materials to surface coatings. In: Laser Surface Engineering; Processes and Applications. Ch. 16 ed. Witney, Oxford, UK: Woodhead Publishing; 2015. pp. 405-422

[31] Hong J-J, Yeh W-C. Application of response surface methodology to establish friction model of upset forging. Advances in Mechanical Engineering. 2018;**10**(3):1687814018766744

[32] Abdul Wasy Zia AW, Zhou ZF, Kwok-Yan LL. A preliminary wear studies

of isolated carbon particles embedded diamond-like carbon coatings. Tribology Internatinal. 2017;**114**:42-47

[33] Yassin MA, Khairul SMS, Mahadzir I, Basim AK. Titanium and its alloys. International Journal of Scientific Research. 2014;**3**(10):1351-1361

[34] Tada H. Manufacturing technologies for titanium eyeglass frames. In: Titanium for Consumer Applications. Ch. 13 ed. New York, USA: Elsevier; 2019. pp. 235-268

[35] Feuerhack A, Trauth D, Mattfeld P, Klocke F. Fine Blanking of Helical Gears. 60 Excellent Inventions in Metal Forming. New York, USA: Elsevier; 2015. pp. 187-192

[36] Zheng Q, Zhuang X, Gao Z, Guan M, Ding Z, Hong Y, et al. Investigation on wear-induced edge passivation of fine-blanking punch. The International Journal of Advanced Manufacturing Technology. 2019;**104**:4129-4414

[37] Aizawa T, Fuchiwaki K, Dohda K. Galling-free fine blanking of titanium plates by carbon supersaturated tool-steel punch. Materials Research Proceedings. 2023;**28**:869-878

[38] Aizawa T, Fuchiwaki K. Galling-free fine blanking of titanium gears using the carbon-supersaturated YXR7 punch. In: Proc. 15th ICTP Conference. Cannes, France: LMNE; 17 Sep 2023. pp. 81-88

[39] Aizawa T, Fuchiwaki K. Galling-free fine blanking of titanium plates using carbon supersaturated high-speed steel punch. Journal of Carbon Research. 2023;**8**(15):1-12

[40] Kuniaki DK, Boher C, Rezai-Aria F, Mahayotsanun N. Tribology in metal forming at elevated temperatures. Friction. 2015;**3**(1):1-27

[41] Funazuka T, Dohda K, Mahayotsanun N. Applications of DLC coating to metal forming. In: Diamond like Carbon Coating: Technologies and Applications. Ch. 7 ed. Boca Raton, FL, USA: CRC Press; 2023. pp. 197-220

[42] Dubois A, Filali O, Dubar L. Effect of roughness, contact pressure and lubrication on the onset of galling of the 6082-aluminum alloy in cold forming, a numerical approach. Wear. 2024;**536-537**:205179

[43] Volz S, Launhardt J, Bay N, Hu C, Moreau P, Dubar L, et al. International round robin test of environmentally benign lubricants for cold forging. CIRP Annals. 2023;**72**(1):245-250

[44] Moskalioviene T, Galdikas A. Stress induced and concentration dependent diffusion of nitrogen in nitrided austenitic stainless steel. Computational Materials Science. 2012;**86**(10):1552-1557

[45] Wang F, Richards VN, Shields SP, Buhro WE. Kinetics and mechanisms of aggregative nanocrystal growth. Chemistry of Materials. 2014;**26**(1):5-21

[46] Wang L, Cao J, Zheng Y, Huang C, Wu S. The theoretical research on technical advance and innovation integration of tapping machinery. IOP Conference Series: Materials Science and Engineering. 2019;**592**:012068

[47] Zhao G-H, Liang XZ, Xu X, Gamża MB, Mao H, Louzguine-Luzgin DV, et al. Alloy design by tailoring phase stability in commercial Ti alloys. Materials Science and Engineering A. 2021;**805**:141229

[48] Pushp P, Dasharath SM, Arati C. Classification and applications of titanium and its alloys. Materials Today: Proceedings. 2022;**54**(2):537-542

[49] Zhang T, Liu C-T. Design of titanium alloys by additive manufacturing: A critical review. Advanced Powder Materials. 2022;**1**:100014

[50] Antolovich SD, Armstrong RW. Plastic strain localization in metals: Origins and consequences. Progress in Materials Science. 2014;**50**:1-160

[51] Aizawa T, Fuchiwaki K. Fine blanking of austenitic stainless steel gears by carbon-supersaturated high-speed steel tools. Machines. 2023;**11**(896):1-13

[52] Domain C, Becquart CS, Foct J. Ab initio study of foreign interstitial atom (C, N) interactions with intrinsic point defects in α-Fe. Physical Review B. 2004;**69**:144122

Chapter 4

Tailoring TiO_2 Films: The Path to Superior Electrochromic Performance

Ayesha Khan and Anamika Vitthal Kadam

Abstract

The present study systematically explores the impact of variations in heating temperature and pH levels on the electrochromic features of electrodeposited TiO_2 thin films. Notably, a TiO_2 film prepared in an acidic medium and annealed at 450°C demonstrated exceptional EC properties. It exhibited a high transmission modulation of 17.18%, an impressive coloration efficiency of 58.8 cm^2/C, and rapid switching kinetics. These results highlight the significance of optimizing synthesis conditions for TiO_2 thin films to enhance their electrochromic performance. The findings of this research contribute valuable insights into tailoring the electrochromic behavior of TiO_2, emphasizing the importance of pH and annealing temperature in achieving enhanced EC properties. The identified optimal conditions for TiO_2 thin films open new avenues for the development of efficient and cost-effective electrochromic materials. This research advances the understanding of TiO_2's electrochromic capabilities and provides a promising candidate for diverse electrochromic applications, thereby expanding the potential impact of electrochromism in various technological fields.

Keywords: TiO_2 films, electrodeposition, indium tin oxide (ITO) glass, pH environment, annealing temperature

1. Introduction

Electrochromism, a fascinating energy conversion phenomenon, has captured the attention of researchers due to its potential applications in diverse fields, including displays, smart windows, antiglare mirrors, and encryption devices [1–4]. It involves the adoption of innovative strategies to reduce energy consumption in buildings, particularly through advanced window technologies, presents significant opportunities. Among these advancements, electrochromism-based smart windows have been developed and utilized in specialized markets for several decades. These windows possess the capability to modify their optical characteristics, such as transmittance and reflectance, with the application of minimal voltage. This distinct feature empowers them to effectively manage solar heat gain, contributing to the establishment of a comfortable indoor environment while optimizing energy utilization [1–4]. The applications of electrochromism-based smart windows are extensive, ranging from automotive and mass transportation to skylights and displays [1–4].

Importantly, the integration of these windows can result in substantial energy savings for buildings. Reports suggest potential reductions of up to 25% in heating and cooling energy consumption, a 50% decrease in lighting energy needs, and a 30–40% decrease in peak power demand [1–4]. Beyond their energy-efficient attributes, these windows also afford privacy protection for occupants [1–4]. The adaptability and demonstrated advantages of electrochromism-based smart windows emphasize their increasing popularity and underscore their pivotal role in advancing sustainable and energy-efficient building technologies across diverse sectors.

It involves the manipulation of electronic bands within chromic materials by an external electric field, resulting in a discernible change in their position and intensity. This change in electronic bands primarily arises from variations in the dipole moment of electronic excitation, with transition polarizability responsible for shifts in intensity [5]. Chromic materials are typically categorized into various assemblies: conducting polymers, small organic molecules, and inorganic metal oxides [6–8]. These EC sources play a crucial role in modulating solar energy entering buildings, contributing to energy conservation by minimizing heating and cooling requirements [9]. Among these materials, TiO_2 has garnered significant attention owing to its non-toxic nature, durability, price-effectiveness, high electrochemical stability, and robust oxidation capabilities [10]. Basically, TiO_2 is a wide bandgap n-type semiconductor with a lattice structure suitable for cation intercalation, leading to the creation of additional electronic states within the material's bandgap. These additional states influence the material's optical properties, enabling light absorption in the visible spectral range [11]. TiO_2 is well-known for its structural polymorphism, featuring phases such as rutile, anatase, and brookite. The control of d-electrons through doping is facilitated by the filling of the 4 s shell, making it more favorable to place the next electron into the 3d shell rather than the 4p shell [12, 13]. TiO_2 naturally exists in a combination of crystalline and amorphous phases [14]. Notably, the amorphous phase of TiO_2 (a-TiO_2) exhibits a large surface area, enhancing catalytic reaction rates and providing opportunities for diverse doping options. Moreover, it is abundant in the environment and can be manufactured at room temperature, eliminating the need for energy-intensive calcination, thus reducing chemical costs and energy consumption [15]. In the amorphous phase of TiO_2, a notable characteristic is the confinement of tail states in close proximity to the band gap, which differs from crystalline materials where tail states are entirely spread out and not confined [14]. These unique attributes of a-TiO_2 have sparked renewed interest in its potential as an alternative to the crystalline phase [15]. Notably, a-TiO_2 has found applications in enhanced photocatalysis [16, 17], dye removal from water [18], and resistive random-access memory [19].

Apart from a-TiO_2, the crystal-like phases of TiO_2, such as anatase and rutile, exhibit unique structural features, chemical attributes, and optoelectronic behavior that set them apart from one another [20–23]. Anatase TiO_2, in particular, has gained widespread recognition as an ideal material for EC applications, primarily due to its open structure, conducive to ion intercalation and deintercalation, resulting in improved EC performance [2]. The transformation of TiO_2 from its natural amorphous or poorly crystalline phases to different crystalline phases involves heat treatment. The phase transition in the material is contingent not only on annealing temperatures but is also squeezed by a multitude of reasons, comprising the reactivity of metal alkoxides, the ratio of water to alkoxides, the pH of the reaction medium, the nature of the solvent employed, and the presence of additives. These combined variables play a crucial role in governing the structural transformation of the material, highlighting the intricate interplay between multiple parameters in the process

[24, 25]. Understanding and controlling these factors are essential for modifying the material's stuffs to meet specific application requirements, making them of significant interest in materials science and engineering [25]. For instance, annealing plays a considerable role in crystal phase formation. The study conducted by Hakki et al. [26] delved into the impact of thermal heating on the crystal structure and crystalline size of TiO_2. The research findings revealed that at a temperature of 500°C, a pure crystalline anatase phase of TiO_2 was formed. However, as the heating temperature was raised to 600°C, this anatase phase disappeared entirely, and only the rutile phase of TiO_2 remained. This transition from anatase to rutile is a significant observation, as these are two distinct crystallographic forms of TiO_2 with varying properties. Furthermore, when the annealing temperature was pushed to a high level of 700°C, the crystalline structure of the rutile phase was destroyed, resulting in TiO_2 being in an amorphous, non-crystalline state. This amorphous state indicates an absence of long-range order in the material's atomic arrangement. The research by Kim et al. [27] explored the influence of heating temperatures fluctuating from 350 to 750°C on both the capacitive and oxidant-generating characteristics of TiO_2. Additionally, this annealing treatment was found to alter the EC properties of TiO_2, as previously indicated in other studies [28, 29]. The pH value was identified as a crucial factor affecting the crystal structures of TiO_2, consequently impacting various properties, including absorption, transmission, and scattering, as reported in previous research [24, 25]. The pH of the precursor solution was detected to enhance the degree of crystallinity, playing a pivotal role in the formation of different phases of TiO_2, a finding consistent with existing literature [30–32]. It is noteworthy that the pH level was found to have a substantial touch on the architectural and photocatalytic features of TiO_2, underlining its importance in shaping the material's characteristics [32]. It was found that the TiO_2 films exhibited a porous structure and significant roughness on their surfaces under lower acidic conditions. In contrast, films were very dense and had relatively smooth surfaces under higher acidic conditions [32]. Consequently, many methods have been adopted to fabricate TiO_2 films with improved EC performance, including the sol-gel technique [33], anodization [11], doctor blade technique [34], vapor deposition [35], hydrothermal [36], and electrochemical deposition [37]. Among them, sol-gel is particularly appealing due to its cost-efficiency and ease of depositing uniform large-area films suitable for window applications [33]. As this method can produce nanocrystalline TiO_2 films, we may expect to see improved EC performance because of its nanocrystalline structure [33]. On the other hand, electrochemical deposition can also be favorable for obtaining TiO_2 thin films since it has low reaction temperatures, flexible substrate selection, and easy scaling [37].

Moreover, Pat et al. [38] emphasized the synthesis of TiO_2 as a highly preferred material using the linear accelerated e-beam evaporation technique to fabricate TiO_2-coated materials on ITO and FTO glass substrates. Electrochemical assessments revealed that TiO_2/ITO and TiO_2/FTO exhibited coloration efficiencies of 26 cm^2/C and 6 cm^2/C, accompanied by achieving ΔT of 10 and 8, respectively. Notably, the TiO_2-deposited layer's optimal contact with ITO-coated glass resulted in superior CE and porosity compared to its deposition on FTO glass. Despite employing the linear accelerated e-beam evaporation technique for TiO_2 fabrication, the low ΔT of 28.6% has rendered the synthesis costly and impractical for real-world applications. In 2020, Nunes et al. [39] reported the synthesis of EC TiO_2 nanostructured films on gold-coated papers using a microwave-assisted hydrothermal method at a low temperature of 80°C. The TiO_2 films, particularly those produced with nitric acid, demonstrated significant ΔT of (57%, 9%, and 22%) between colored and bleached states at 250, 550, and 850 nm, respectively, in reflectance mode. These films also exhibited high cycling stability, enduring up to 1500 cycles without a significant loss of EC behavior. Despite the simplicity of the hydrothermal method, the requirement for gold-coated papers as a substrate for film deposition limits the practical applicability. The use of gold-coated paper does not significantly enhance the EC properties, posing a constraint on the overall effectiveness of the approach. In 2021, Eyovge et al. [40] reported the use of co-axial electrospinning to structure non-woven webs of TiO_2 nanofibers loaded with Ag, Au, and CuO nanoparticles for EC applications. The composite layers exhibited distinct electrochromic effects: TiO_2 loaded with Ag displayed a black-brown color, Au showed a dark-blue color, and CuO showed a dark-green color. Among these, the Au/TiO_2 layer proved the most promising for electrochromic applications, featuring a color modulation time of 6 seconds, transmittance modulation of 40%, coloration efficiency of 20 cm_2/C, areal capacitance of 300 F/cm^2, and cyclic stability of over 1000 cycles. Despite employing metal doping to enhance the EC properties of pristine TiO_2, the study notes that the resulting CE values and stability are not significantly improved. This limitation underscores challenges in achieving substantial advancements even with the introduction of metal dopants, potentially hindering the practical viability of the approach at a commercial scale. In 2022, Almarri et al. [41] reported the synthesis of TiO_2 deposited by dip coating, incorporating poly(3,4-ethylenedioxythiophene)-poly(styrenesulfonate) (PEDOT:PSS) to form TiO_2 + PEDOT:PSS hybrid composite films. EC characterizations demonstrated the reversible color change of the flexible hybrid device from dark blue to white transparent in less than 1 second, with a high in situ transmittance modulation (ΔT = 70%) at 750 nm. However, a notable drawback in the synthesis is the incorporation of organic species along with TiO_2 to enhance EC performance, making the process complex and increasing synthesis costs. In 2023, Xing et al. [42] successfully engineered defect-rich brookite TiO_2 nanorods (NRs) via colloidal approach and examined their electrochromic capabilities for potential energy storage applications. The EC characteristics of TiO_2 NRs, particularly those with longer lengths, demonstrate rapid switching speeds (20 seconds for coloration and 12 seconds for bleaching), high coloration efficiency (84.96 cm^2 C^{-1} at a 600 nm wavelength), and overall stability. While the CE of the produced TiO_2 is notably higher than reported values in the literature, there remains a drawback in terms of the slow switching speed, indicating a need for improvement in this aspect of the electrochromic performance. The drawback of the mentioned synthesis lies in its complexity and potential cost implications due to the controlled engineering of defect-rich brookite TiO_2 nanorods, limiting its practical scalability for widespread applications. In 2023, Aiempanakit et al. [43] synthesized TiO_2 nanotubes (TNTs) and bamboo-type TNTs structure films by anodizing sputtered titanium (Ti) films on ITO glass, adjusting the anodization process with varying amounts of deionized water and ethylene glycol. The EC performance was influenced by the amorphous structure and high energy band gap (Eg) in TNTs before annealing, with the bamboo-like structure exhibiting the best electrochromic properties, achieving ΔT of 12.58%. Despite utilizing the sputtering and anodization method for the synthesis of TiO_2 nanotubes and bamboo-type TNTs structure, the achieved low transmission modulation of only 12.58% renders it impractical for economic scale applications. Additionally, the time-consuming nature of the process further adds to its drawbacks. Additionally, the studies [26–29, 30–32, 38–43] demonstrate the annealing effect and pH on the EC properties of TiO_2 deposited by different methods. Additionally,the thickness of the TiO_2 film can be efficiently controlled by controlling

Sr. no	Sample	Method	EC properties					Ref
			Optical modulation (ΔT%)	CE (cm^2/C)	Response time (sec)		Stability (CV cycles)	
					tb	tc		
1	TiO_2	Anodization	—	—	30	30	50	11
2	TiO_2	Chemical solution deposition	14.2	7.98	30	30	—	28
3	TiO_2	Sol-gel method	—	—	300	300	—	33
4	TiO_2	Doctor blade method	68	33.7	2.0	5.0	—	34
5	TiO_2	hot filament metal oxide vapor deposition	67	226	13.0	5.4	2000	35
6	TiO_2	electrodeposition	—	—	—	—	1500	37
7	TiO_2	linear accelerated e-beam evaporation technique	10	26	—	—	—	38
8	TiO_2	hydrothermal method	9.0	—	58	71	1500	39
9	TiO_2	Electrospinning	5	2.31	9.6	10.8	1000	40
10	TiO_2	Colloidal approach	66.87	84.96	20	12	1000	42
11	TiO_2	Anodization	12.58	—	—	—	—	43
12	TiO_2	electrodeposition	17.18	58.33	1.32	2.10	1500	Our work

Table 1.
EC properties of TiO_2 reported in the literature.

the applied current voltage and deposition time [37]. The EC properties of TiO_2 reported in the literature have been described in **Table 1**.

Nevertheless, the influences of thickness, annealing temperatures, and pH on the EC activities of TiO_2 via simple electrodeposition using PG ($C_3H_8O_2$) have not been extensively explored. Therefore, the current investigation goals to sermon this knowledge gap by directing an orderly examination into the effects of thickness, annealing temperature, pH on the EC properties of electrodeposited TiO_2.Till date, there is currently no comprehensive study that has systematically examined the combined impact of these parameters, including annealing temperature, and pH on the EC properties of TiO_2 thin films. This research gap signifies the novelty and significance of the present study, as it seeks to shed light on how these factors collectively influence the EC behavior of TiO_2 films. Furthermore, the utilization of TTIP as a precursor and PG as both a solvent and a binder present a promising approach for producing stable films at a reduced cost, which is an innovative aspect of this research that contributes to the advancement of materials synthesis and their practical applications.

2. Experimental

The experimental exploration was segmented into Two specific components: *Section A (variation in annealing temperature), and Section B (variation in pH levels).*

2.1 Materials

The materials utilized in the work included Titanium isopropoxide (TTIP), $C_3H_8O_2$, HCl, DDW, Ammonia (NH_3), and ITO glass plates with a conductivity of 25 ohm-cm^2 and dimensions of 3 × 1 cm^2, employed as substrates.

2.2 Synthesis of TiO_2 films

In this study, TiO_2 films were synthesized through an electrodeposition method on ITO glass substrates. Before commencing the synthesis process, the ITO glass substrates underwent a rigorous cleaning procedure involving ultrasonication in distilled water and acetone for a duration of 15 minutes. All aqueous solutions utilized in the experiments were prepared meticulously using DDW. The primary objective of this synthesis was to precisely tailor and optimize the fabrication process of TiO_2 films, with a specific emphasis on their suitability for EC applications. The synthesis procedure was initiated with systematic variations in heating temperatures within the range of 250–650°C (*SECTION A*). After comprehensive evaluation, it was monitored that an annealing temperature of 450°C yielded the most favorable anatase phase, resulting in the best EC application of TiO_2. Therefore, this temperature was selected for further synthesis steps. Lastly, the pH of the precursor solution was subjected to optimization, ranging from acidic to basic. Through experimentation, it was found that an acidic pH level of the precursor solution facilitated optimal growth of TiO_2 crystals with improved EC performance. Consequently, this pH level was chosen as the most suitable for the synthesis process (*SECTION B*). The synthesis process of TiO_2 films commenced by preparing a 0.05 M solution through the vigorous mixing of TTIP with PG. To attain the target pH of 1.0, HCl was cautiously introduced in small drops to get the clear TTIP solution. The reaction ensued as TTIP and PG formed a complex, subsequently undergoing hydrolysis. This chemical transformation led to the production of TiO_2 and propionaldehyde, marking a critical step in the fabrication of TiO_2 films. In this reaction, PG served as both a solvent and a stabilizer, preventing the formation of unwanted byproducts (Eq. (1)).

$$Ti\left[OCH\left(CH_3\right)_2\right]_4 + 2CH_3CH\left(OH\right)CH_2OH \rightarrow TiO_2 + 4CH_3CHO + 4H_2O \quad (1)$$

TiO_2 thin films were fabricated through the deposition process, utilizing a three-electrode electrochemical cell configuration. In this setup, graphite assisted as the counter electrode, an Ag/AgCl electrode was used as the reference electrode, and ITO was employed as the working electrode. The deposition of TiO_2 thin films took place at ambient room temperature, with a consistent voltage of −0.4 V applied for a duration of 1000 seconds.

For *SECTION A*, after deposition, the TiO_2 film was dried at room temperature and annealed at temperatures of 250°C, 450°C, and 650°C to induce crystallization, and they were designated as T_{250}, T_{450}, and T_{650}, respectively. For *SECTION B*, HCl and NH_4OH were used to adjust the pH of the precursor solution to pH values of 1, 6, and 10. After deposition, the TiO_2 film was dried at ambient temperature and annealed at 450°C to induce crystallization, and they were named T_A (pH = 1), T_N (pH = 6), and T_B (pH = 10).

2.3 Methods

For the identification of the crystal phase and to calculate various structural parameters of the obtained sample, XRD patterns were recorded using XPERT-PROMPD X-ray diffractometer with uKα radiation (λ = 1.5405 Å) in 2θ range of 10–80°. The surface morphology of the synthesized products was observed using FE-SEM (Gemini attached in EDS, Oxford, AZtecLive).). The FT-IR spectrum was recorded in the range of 500–4000 cm^{-1} at a resolution of 2 cm^{-1} using the Perkin–Elmer 1710 spectrophotometer. The absorbance spectra were obtained using a UV-Vis spectrophotometer (PerkinElmer, Lambda 750). The film thickness ranged from 360 to 430 nm, and was determined using a surface profiler (DEKTAK3ST, Veeco). For the purpose of revealing the chemical characteristics, X-ray photoelectron spectroscopy (XPS) (Thermo Fischer Scientific ESCALAB Xi+) was used. The electrochemical features like CV, CA, and EIS were performed using a three-electrode system (Electrochemical Analyzer 608, CH Instruments), the counter Electrode (Platinum electrode), RE (Ag/AgCl electrode) and a WE (TiO_2 thin film) with an electrolyte solution containing 0.5 M $LiClO_4$-PC.

3. Section A (variation in annealing temperature)

3.1 Results and discussions

3.1.1 XRD analysis

The crystalline phase of TiO_2 films was studied using XRD measurements. The TiO_2 material displayed a crystalline structure in both T_{250} and T_{450} films, evidenced by a distinctive diffraction peak observed at 2θ values of 25.3°, corresponds to the (101) Miller index corroborating the anatase phase (JCPDS files No. 21–1272) (**Figure 1**) [44, 45]. The (*) marked indicates substrate (ITO) peaks. It was perceived that while the temperature amplified from 250 to 450°C, the intensity of the anatase peak (101) was increased, implying an improvement of the crystalline structure and hence an increase in the crystallite size [46]. Another feature prevalent in

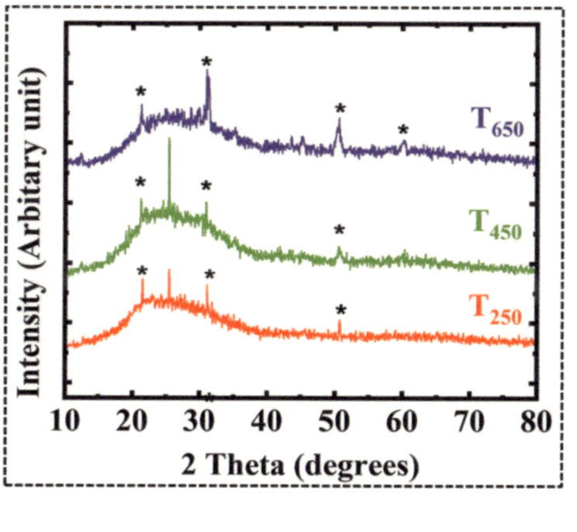

Figure 1.
XRD analysis of annealed TiO_2 films.

Sample	Nature	Crystalline size (nm)
T_{250}	amorphous+crystalline	5.2
T_{450}	amorphous+crystalline	10.4
T_{650}	amorphous	—

Table 2.
Crystalline size of annealed TiO2 films.

diffractograms is the presence of a crystalline phase along with an amorphous phase associated with efficient ion intercalation/deintercalation, resulting in enhanced EC performance [47]. It was found that annealing at 650°C causes anatase to revert to an amorphous form, like in previous reports [26]. The average grain size is computed using Scherrer's equation. **Table 2** shows the crystallite sizes calculated, which range from 5.0 to 11 nm depending on the temperature of the thermal action. In our scenario, we noted that the point at which crystallinity becomes apparent in the films occurs at a minimal temperature, specifically 250°C. The variation in initiation temperatures for crystallization is directly associated with the specific precursor materials employed [48]. The XRD results of the TiO_2 film developed for EC applications reveal the existence of anatase as well as brookite phases of TiO_2, each possessing well-established electrochromic properties [38–45]. Despite the presence of these multiple phases, there is a clear inclination in the literature to preferentially utilize the anatase phase of TiO_2 in films designed for EC applications. This preference is rooted in the open structure of the anatase phase, which facilitates efficient ion intercalation and deintercalation, ultimately leading to an improved overall EC performance [2].

3.1.2 FE-SEM analysis

Figure 2 illustrates characteristic FE-SEM images of TiO_2 explored as a function of temperature. A change in the annealing temperature, apart from variations in crystalline structure observed in XRD, is also responsible for significant changes in the morphology of TiO_2 film.. It can be seen that annealed TiO_2 displays a very cracked morphology similar to the crack morphology of nanocrystalline Titania [49], typical of the annealed film where cracks are formed when capillary stress exceeds the tensile stress of the film [49]. It is also observed that cracks spread entirely throughout the film, referred to as channeling cracks. Besides the damage, it also exhibits portions of agglomerated points (**Figure 2(d)**), similar to those observed by Leftheriotis and Yianoulisa [50]. The crack morphology observed in TiO_2 films facilitates the easy penetration of electrolyte ions within the film, making the film more favorable for efficient ion diffusion during intercalation and deintercalation [51]. The present study may conclude that the crack morphology is obtained due to the tensile stress applied by a more viscous solvent (PG) on TiO_2 nanoparticles. Additionally, TiO_2 films form a uniform and densely packed structure with a proliferation in heating treatment from 250 to 650°C, implying a reduction in crack spacing and a decline in EC performance due to insufficient ion percolation.

3.1.3 FTIR analysis

Chemical bonding between Ti and O_2 atoms in annealed TiO_2 films was scrutinized through the utilization of FT-IR spectroscopy.

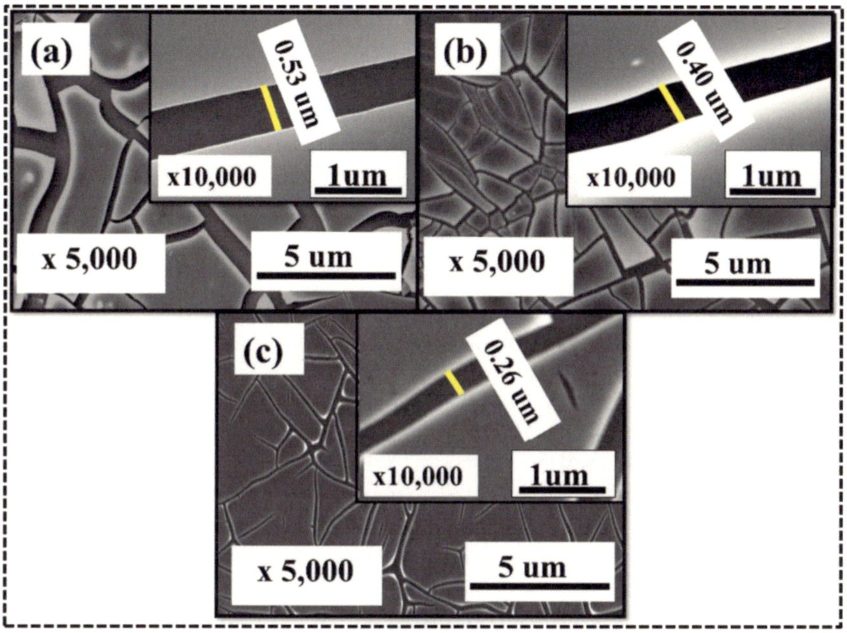

Figure 2.
FE-SEM analysis of annealed TiO_2 films at (a) 250°C (b) 450°C (c) 650°C.

Distinctive vibrational modes of the synthesized TiO_2 material were identified at wavenumbers of 417 cm^{-1}, 520 cm^{-1}, and 870 cm^{-1}, and these can be credited to the stretching vibrations within the inorganic Ti–O–Ti network (**Figure 3**) [52, 53]. Furthermore, a band at 3350 cm^{-1} is ascribed to the stretching vibrations associated with hydroxyl groups (Ti-OH) originating from residual alcohol and hydroxyl groups [52, 53]. It was noteworthy that the annealing treatment resulted in a reduction in the intensity of the stretching vibration at 870 cm^{-1}. This phenomenon is attributed to the existence of both the anatase and amorphous phases of TiO_2 with increasing annealing

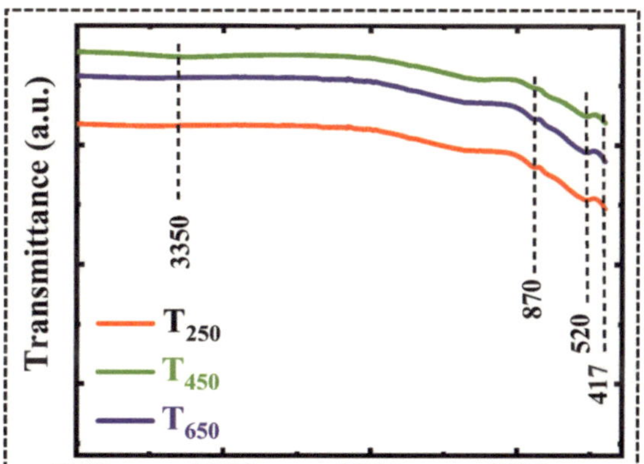

Figure 3.
FTIR analysis of annealed TiO_2 films.

temperature. In its amorphous state, TiO_2 typically exhibits a characteristic signal around 870 cm^{-1} [54]. The diminished intensity of the stretching vibration at 870 cm^{-1} suggests the existence of nanograins of the anatase phase, even within the amorphous TiO_2 films. Importantly, the absence of other absorption peaks, such as those associated with organic residues like –CH and –CH_2 stretching vibrations in the range of 1400–2900 cm^{-1}, indicates the absence of carbon contamination in the films. These findings align with the appearances of the anatase phase of TiO_2 as documented in existing literature [52, 53].

3.1.4 EC and electrochemical analysis

3.1.4.1 CV analysis

The EC investigation of annealed TiO_2 films was carried out through CV analysis in the potential array spanning from −2 to 1 V, with reference to an Ag/AgCl electrode, at an examination rate of 50 mV/s, as depicted in **Figure 4**.

During the positive cycle of CV, it was observed that the annealed TiO_2 films, particularly the T_2 film, displayed redox peaks occurring at approximately −0.73 V, characteristic of the anatase phase of TiO_2. Notably, these redox peaks exhibited slight shifts with increasing annealing temperature, possibly attributed to the heightened internal diffusion resistance within the film. The CV curve of the TiO_2 film displays a distinctive duck-shaped profile, closely resembling the CV curves documented in previous literature [39, 40]. The T_{450} film has demonstrated a higher current density, which is indicated by the larger area under the CV curve. This improvement in performance can be attributed to the increased diffusion area owing to the crack morphology that the film offers to the electrolyte, surpassing the performance of other films. This underscores the beneficial impact of larger electrode surface areas on Li intercalation and deintercalation processes, aligning with the principles of the Randles-Sevcik equation [40, 55, 56].

3.1.4.2 CA analysis

It is of paramount importance to accurately assess the period mandatory for the anodic or cathodic current to attain a stable level following the application

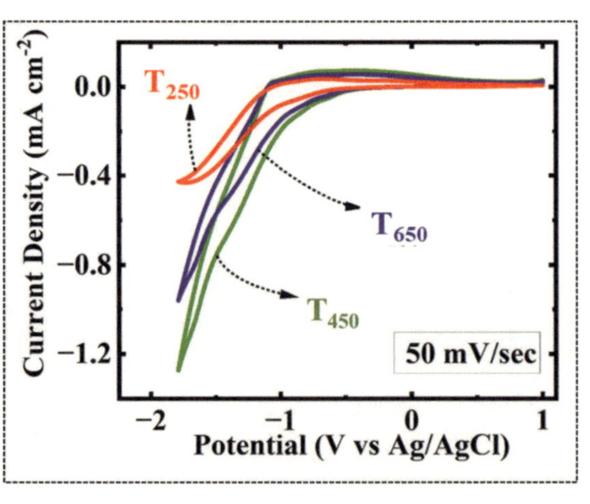

Figure 4.
CV analysis of annealed TiO_2 films.

of responsive voltages, particularly when transitioning between the colored and bleached states [57]. In our EC investigation, the amperometric i-t curve was instrumental in characterizing the switching kinetics of TiO_2 films subjected to annealing at various temperatures. These films were immersed in a 0.5 M $LiClO_4$-PC electrolyte for a duration of 20 seconds, as illustrated in **Figure 5**. The determination of response times for both the colored and bleached states of the TiO_2 films is presented in **Table 3**. The switching kinetics of the coloration reaction in the T_{450} film is faster than in the T_{250} and T_{650} films. The quicker response of T_{450} originates from its island morphology, which facilitates easy and fast diffusion of ions within the film and improves the transmittance contrast and switching speed simultaneously.

3.1.4.3 Optical analysis

The annealing treatment also affects the transmittance of TiO_2 films. The optical transmittance of WO_3, TiO_2, and WT films was assessed using a spectrophotometer in both their colored and bleached states within a 0.5 M $LiClO_4$-PC electrolyte (**Figure 6**). The change in optical density (ΔOD) was calculated using the following equation (Eq.2) [8]:

$$\Delta OD_{\lambda=630\,nm} = \log\left(\frac{T_b}{T_c}\right) \quad (2)$$

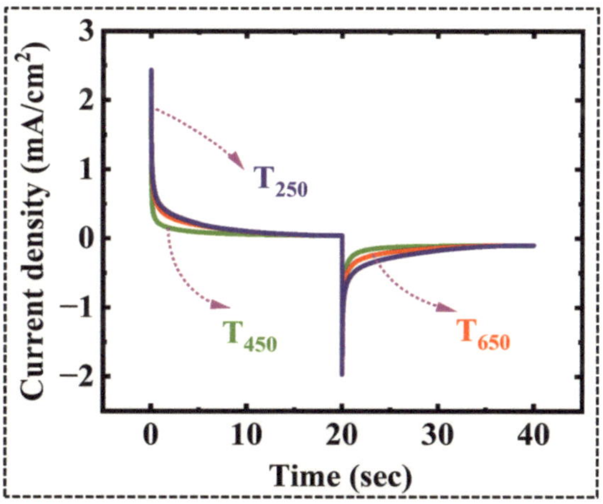

Figure 5.
CA analysis of annealed TiO_2 films.

Sample	Response time(sec)		%T_b	%T_c	%Δ T	$(\Delta OD)_{630nm}$	C.E (cm^2/C)
	t_c	t_b					
T_{250}	4.23	5.43	83.5	70.2	13.3	0.063	36.41
T_{450}	1.32	2.10	82.3	64.5	17.18	0.010	58.83
T_{650}	3.28	4.25	81.1	66.9	14.2	0.083	48.14

Table 3.
EC properties of annealed TiO2 films.

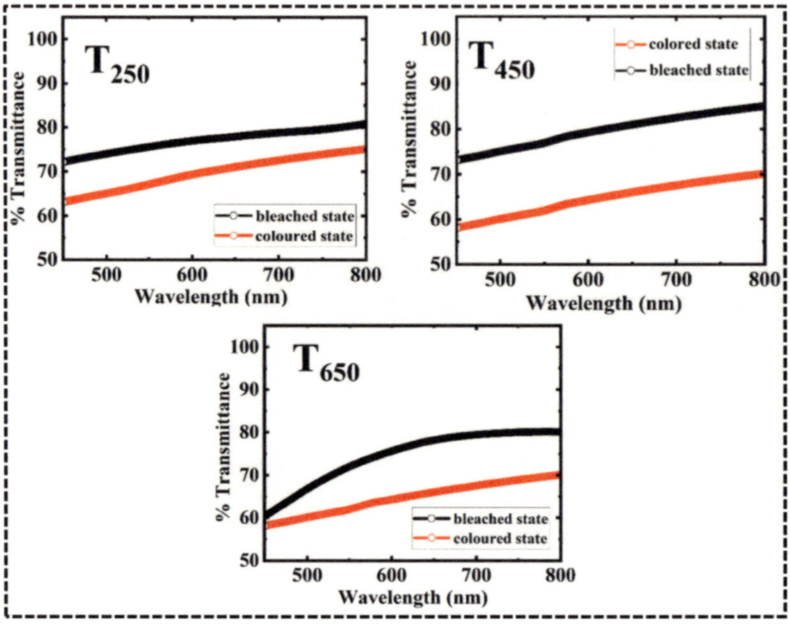

Figure 6.
The transmittance spectra of annealed TiO_2 films in their colored and bleached states.

T_b and T_c are the transmittance values for the bleached and colored states at 630 nm [8].

The CE, a vital parameter to probe the potential of the material as an EC material, was calculated at 630 nm using the following relation (Eq. 3) [8]:

$$CE_{\lambda=630nm} = \frac{\Delta OD_{630nm}}{Q_i} \quad (3)$$

Where $\Delta OD_{630nm} = \log\left(\frac{T_b}{T_c}\right)$ And Qi = charge intercalated.

The T_{450} film in the current work exhibits the highest transmission modulation of 17.81% with CE of 58.38cm^2/C at 630 nm. The main reason behind this is the enhanced ion and electron diffusion and can be attributed to the abundance of cracks and crevices within T_{450} film. After annealing the TiO_2 film to a higher temperature (650 °C), CE decreased as ions and electrons could not intercalate and de-intercalate efficiently as the film tended to become smooth. Hence, the optimized annealing temperature of 450 °C has been chosen for the pH series.

4. Section B (variation in pH)

4.1 Results and discussions

4.1.1 XRD analysis

The XRD statistics of TiO_2 films exhibited notable variations in response to modifications in the pH of the pioneer solution. **Figure 7** provides an insightful

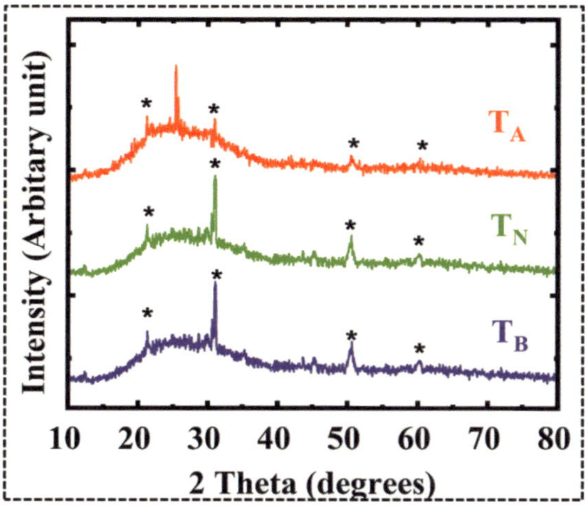

Figure 7.
XRD records of TiO_2 films in different pH media.

depiction of the crystal structure and phase purity of TiO_2 thin films manufactured under different pH conditions (pH 1, 6, and 10). Upon examination of the XRD profiles, it is evident both TA and TN films prominently display characteristic peaks of anatase phase at 2θ values of 25.3° corresponding to the (101) crystal planes (JCPDS files No. 21–1272) [44, 45]. Notably, the peaks marked with asterisks (*) represent the substrate (ITO) peaks. A discernible trend is observed in which the peak intensity diminishes with increasing pH values, indicating a reduction in crystalline size. The crystallite size of TiO_2 films was quantified using Scherrer's equation (**Table 4**). It is documented that the crystallite size can be effectively modulated by adjusting the pH of the precursor solution [58]. The reduction in crystallite size with rising pH can be ascribed to the repulsive forces operating between particles, preventing the coalescence of nanocrystalline entities, and impeding particle growth [59]. Furthermore, the existence of the anatase phase in the TA film, featuring a prominent peak at (110), signifies that the acidic environment fosters a more pronounced crystal growth compared to other pH conditions.

4.1.2 FE-SEM analysis

Figure 8 exhibits characteristic FE-SEM images of TiO_2 investigated as a function of pH of precursor solution. As can be seen, TiO_2 films exhibit channeling cracks, which are characterized by island-like crack patterns. It was observed that islands at higher pH have much larger sizes than islands at lower pH. As established by numerous scientific investigations, the surface charge of TiO_2 is intrinsically induced by

Sample	Nature	Crystalline size (nm)
T_A	amorphous+crystalline	10.4
T_N	amorphous+crystalline	3.5
T_B	amorphous	—

Table 4.
Crystalline size of TiO_2 films in different pH media.

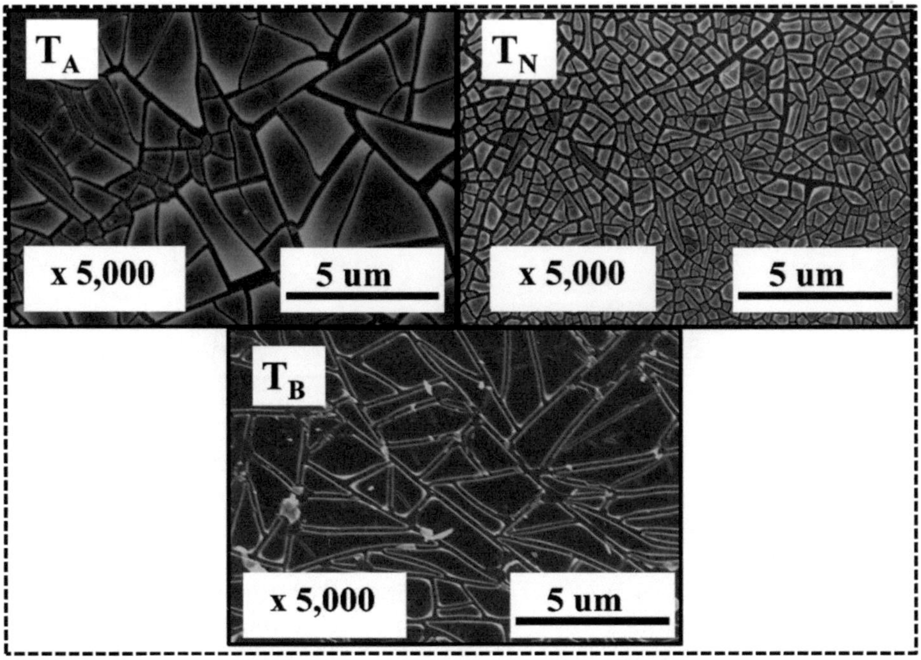

Figure 8.
FE-SEM analysis of TiO_2 films in different pH media.

variations in pH [30]. Under acidic conditions, the presence of a robust repulsive charge among TiO_2 particles diminishes the likelihood of coalescence. This phenomenon results in the creation of a more stable sol, ultimately leading to the development of cracks with larger dimensions. These pronounced cracks, in turn, facilitate the growth process [30]. According to FE-SEM images (**Figure 8**), growth rate of TiO_2 decreases as pH increases from acidic to basic. The crack morphology observed in TA enables the easy penetration of electrolyte ions within the film, making the film more favorable for efficient ion diffusion during intercalation and deintercalation, as confirmed by electrochemical analysis [49–51].

4.1.3 FTIR analysis

FTIR stands as a practical and indispensable technique for the recognition of functional groups present in prepared samples.

Across the pH range from acidic to basic, the FTIR spectra reveal distinctive vibrational modes associated with the synthesized TiO_2 material, specifically at wavenumbers of 404 cm^{-1}, 546 cm^{-1}, and 799 cm^{-1} (**Figure 9**). These vibrational modes can be ascribed to the stretching vibrations inherent to the inorganic Ti–O–Ti network [52, 53]. Additionally, the presences of a characteristic peak at 1324 cm^{-1} is noteworthy, and its broadness is more pronounced at higher pH levels. This peak can be accredited to both C–H bending and C–O stretching modes of vibration. Furthermore, a band at 3451 cm^{-1} is ascribed to the stretching vibrations associated with hydroxyl groups (Ti-OH) originating from residual alcohol and hydroxyl groups [52, 53]. These observations align with the characteristics of the anatase phase of TiO_2, as reported in the existing literature [52, 53].

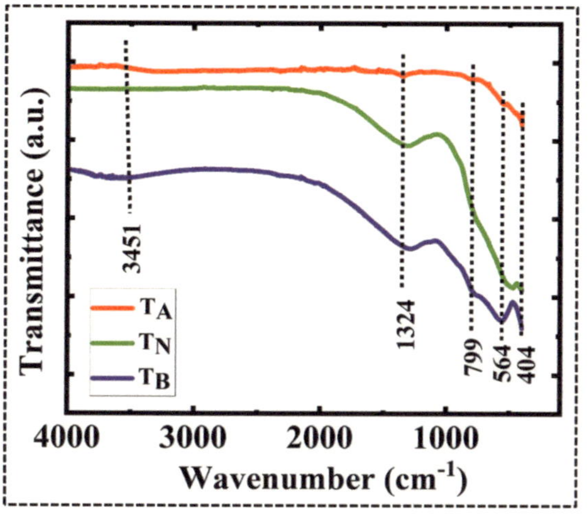

Figure 9.
FTIR analysis of TiO_2 films synthesized in different pH media.

4.1.4 XPS analysis

The investigation utilized XPS to explore the elemental composition and chemical states of TiO_2 across various pH conditions, as depicted in **Figure 10(a–f)**.

Analysis of the XPS spectrum of TiO_2 films (**Figure 10**) revealed the exclusive presence of titanium (Ti) and oxygen (O) elements. In the T_A sample (**Figure 10(a)**), distinctive peaks were identified at binding energies (BE) of 458.5 eV and 464.2 eV. Similarly, the T_N sample (**Figure 10(c)**), exhibited notable peaks at BE 458.7 eV and 464.3 eV, while the T_B sample (**Figure 10(e)**), displayed peaks at BE 458.9 eV

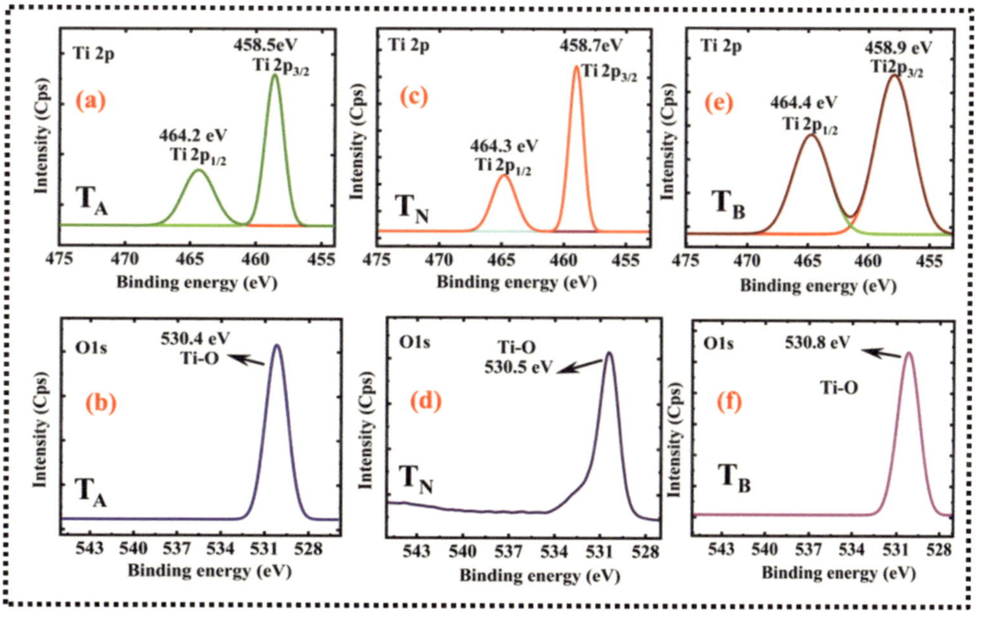

Figure 10.
XPS analysis of (a-b) T_A (c-d) T_N (e-f) T_B films.

and 464.8 eV. These peaks denote the presence of $Ti2p_{3/2}$ and $Ti2p_{1/2}$, distinctly characteristic of Ti^{4+} ions incorporated within the TiO_2 lattice [60, 61]. Moreover, the computed difference in BE between $Ti2p_{3/2}$ and $Ti2p_{1/2}$ ($\Delta BE = Ti2p_{3/2} - Ti2p_{1/2}$) for the T_A sample was approximately 5.7 [60, 61]. Similarly, for the T_B sample, ΔBE was around 5.6, and for the T_N sample, it was about 5.5, consistent with the characteristic Ti^{4+}-O bonds in TiO_2 [60, 61]. Additionally, examination of the O_{1s} spectra in the T_A sample (**Figure 10(b)**), revealed a singular signal at BE 530.4 eV, while the TN sample (**Figure 10(d)**), displayed a peak centered at BE 530.5 eV, and T_B exhibited a signal at BE 530.8 eV (**Figure 10(f)**). These signals are attributed to lattice oxygen (O) within the Ti-O bond. The calculated ΔBE between Ti_{2p} and O_{1s} was 71.9 for T_A, 71.8 for T_N, and 71.9 for T_N samples, indicating a proximity to the anatase phase (71.4 eV). These findings are consistent with the existing literature [60, 61].

4.1.5 EC and electrochemical properties

4.1.5.1 CV analysis

The electrochemical assessment of the TiO_2 films was conventionally conducted through CV, operating within a potential range encompassing the 0.5 M $LiClO_4$-PC electrolyte. As presented in **Figure 11**, the CVs were systematically scanned within a potential span extending from −2 to 1.0 V at a scan rate of 50 mV/s, with reference to Ag/AgCl electrodes. For metal oxides such as TiO_2, the process of ion intercalation corresponds to optical coloration, while deintercalation is linked to the bleaching process, both of which are extracted in terms of the diffusion coefficient (D) [62]. It can be quantified through the utilization of the Randles-Sevcik equation [Eq. (4)].

$$ip = 2.72 \times 105\frac{\frac{n3}{2}D1}{2}CO\frac{\nu 1}{2} \quad (4)$$

Where D is the diffusion coefficient of Li + ions; ν, scan rate (0.05 V/s); C0, concentration of active ions in the solution (0.5); n, number of electrons (it is assumed to

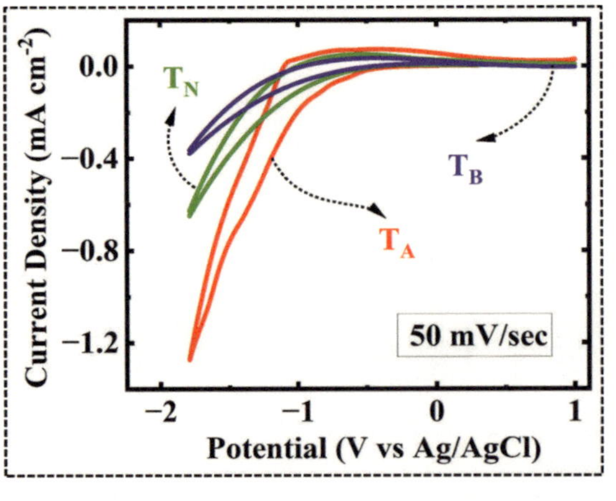

Figure 11.
CV analysis of TiO_2 films synthesized in different pH media.

Sample	Ip (A)	Diffusion coefficient (cm^2/s)
T_A	0.0032	$1.743*10^{-8}$
T_N	0.0012	$1.568*10^{-9}$
T_B	0.0005	$2.723*10^{-10}$

Table 5.
Diffusion coefficients of TiO_2 films in different pH media.

be 1) and ip is the peak current density [62]. The diffusion coefficients are calculated for different pH media and are presented in **Table 5.**

We have noticed that the current density, represented by the area under the curve, for the T_A electrode exceeds that of both T_B and T_N electrodes at a scan rate of 50 mV/s. This disparity can be attributed to the larger diffusion area available for the acidic medium, resulting in a more effective interaction with the electrolyte compared to the basic and neutral states. Additionally, it is evident that the redox peaks of the TiO_2 films exhibit slight shifts towards higher potentials as the pH value increases. This phenomenon is likely associated with an escalation in the internal diffusion resistance within the film [35]. It is noteworthy that in an acidic environment with a pH of 1, the TiO_2 film demonstrates an expanded surface area, which enables the accommodation of a large number of active sites for the intercalation and de-intercalation of Li^+ ions. This characteristic is particularly advantageous in EC applications.

4.1.5.2 CA analysis

The assessment of the duration vital for the anodic or cathodic current to reach a stable state is a critical parameter, especially when reactive voltages are applied to transition between the colored and bleached states, as it directly impacts the reliability and effectiveness of EC films [57]. In our EC investigation, we employed amperometric i-t curves to assess the response time of TiO_2 films under varying pH conditions, with a dwell time of 20 seconds in a 0.5 M $LiClO_4$–PC electrolyte (**Figure 12**). The calculated response times for achieving both the colored and decolored states of the TiO_2 films are presented in **Table 6**. Our observations indicate that at lower pH values, characterized by a higher content of H+ ions, a substantial number of small nuclei are produced, possessing a notably enlarged active surface area [32]. In the current study, we found that the switching kinetics governing the coloration reaction in the T_A film are faster in comparison to the T_B and T_C films. This boosted reactivity can be credited to the amplified insertion of H^+ ions at pH = 1 [32]. This effect is closely associated with the presence of a larger active specific superficial area and the distinctive morphology inherent to crystalline TiO_2 nanostructures. In this investigation, the observed switching speeds, particularly the bleaching time of 2.10 seconds, were marginally faster compared to values reported in previous studies (**Table 1**) [11, 28, 33–43]. Notably, these switching speeds are better than those documented in earlier literature [11, 28, 33–43]. The enhanced switching speed can be ascribed to the island-like morphology illustrated in **Figure 5**. This morphology facilitates the swift and effortless diffusion of ions within the film, leading to simultaneous improvements in transmittance contrast and switching speed.

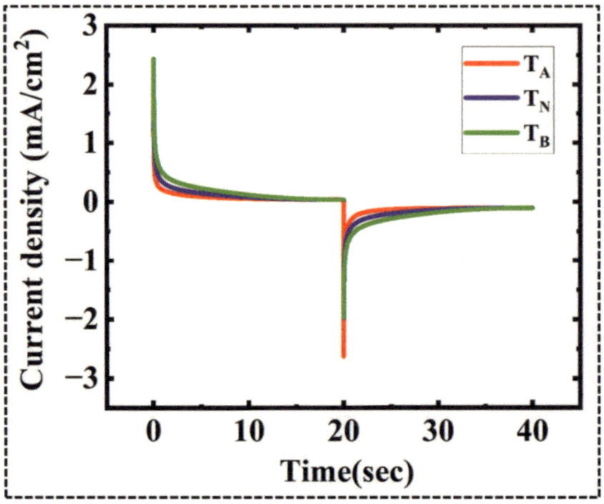

Figure 12.
CA analysis of TiO_2 films synthesized in different pH media.

Sample	Response time (sec)		$\%T_b$	$\%T_c$	$\Delta\%T$	$(\Delta OD)_{630nm}$	CE (cm^2/C)
	t_c	t_b					
T_A	1.32	2.10	82.3	64.5	17.18	0.010	58.38
T_N	2.66	3.20	82.3	70.8	11.5	0.052	31.07
T_B	3.78	4.04	83.1	75.8	7.3	0.039	29.35

Table 6.
EC properties of TiO_2 films in different pH media.

4.1.5.3 Optical analysis

In order to estimate the effect of pH on EC response time, 95% of maximum optical contrast at 630 nm was considered. The study examined the transmittance of TiO_2 films in both their colored and bleached states as depicted in **Figure 13**. To quantify the change in ΔOD, Eq. (2) was employed. Furthermore, the efficiency of the coloration process, denoted as CE and considered a crucial factor in assessing the material's suitability for EC applications, was determined at a wavelength of 630 nm using Eq. (3). The investigation revealed a noticeable reduction in ΔOD as the pH increased, consequently leading to a decline in the CE of the TiO_2 films. Notably, within the scope of the present work, the TA film exhibited the most substantial optical density, reaching 0.065 at 630 nm when transitioning between the bleached and colored states. This heightened ΔOD is attributed to the crystalline nature of the T_A nanostructures, which possess an extensive active surface area and a favorable crack morphology. As a result, the acidic medium is established as advantageous for the development of the anatase phase, thereby enhancing the EC properties, particularly the noteworthy improvement in CE compared to other crystalline systems [51]. In the present study, the T_A film demonstrates the highest percentage change in transmittance modulation of 17.18%, coupled with a remarkable CE charge value of 58.38 cm^2/C at a wavelength of 630 nm. Additionally, the

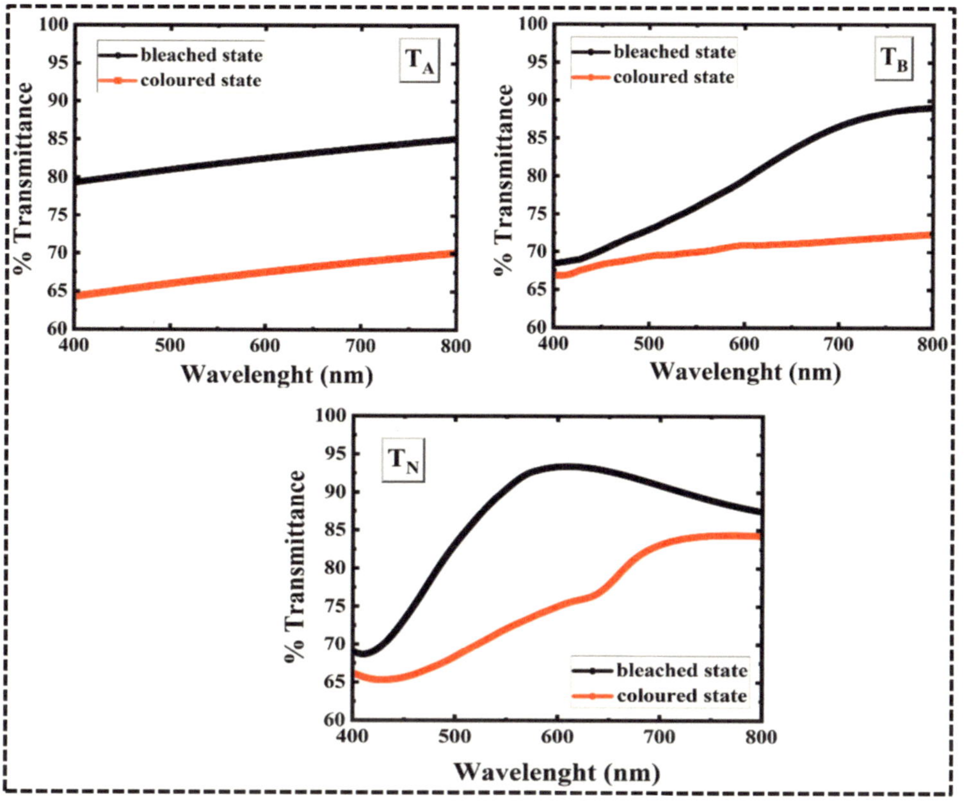

Figure 13.
The transmittance spectra of the TiO_2 films in their colored and bleached states.

CE value for the T_A film surpasses those documented in previous literature [28, 34, 38, 40], reaching levels comparable to CE values attained through alternative methodologies [42].

Consequently, the crystalline TiO_2 film produced in this study holds significant potential for application in energy-efficient smart windows, owing to its swift-switching response and commendable CE. The EC parameters of the synthesized TiO_2 films are comprehensively outlined in **Table 6.**

4.1.5.4 Stability

In the realm of EC materials, assessing stability is a pivotal criterion for evaluating their overall performance and suitability for various applications. In a current study, we undertook comprehensive stability testing on a TiO_2 film, employing a 0.5 M $LiClO_4$-PC electrolyte. The testing was carried out over a defined potential range of −2.0 to +1.0 V versus an Ag/AgCl (**Figure 14**) [51]. The results of our investigation revealed a remarkable outcome: the TiO_2 film exhibited a level of stability that greatly surpassed expectations. Specifically, it demonstrated an impressive capacity to endure approximately 1500 charge and discharge cycles without significant deterioration or loss of functionality. This achievement exceeds the performance documented in existing literature for TiO_2 films, as reported in literature [11, 40, 42], and is on par with the stability attained through alternative methods [37, 39]. This enhanced stability

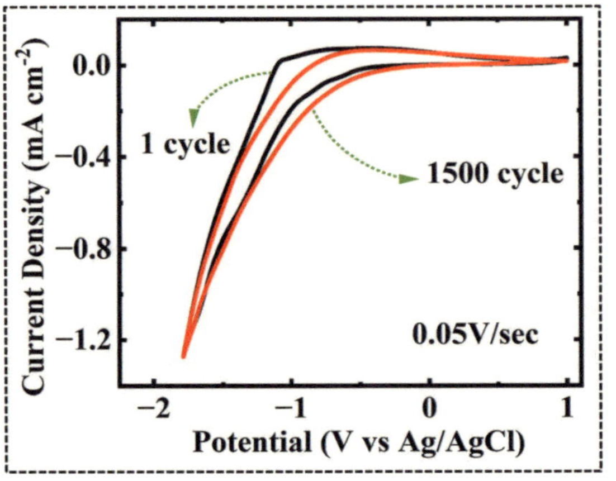

Figure 14.
CVs recorded for T_A film in 0.5 M $LiClO_4$-PC electrolyte.

underscores the potential of the TiO_2 film as a promising candidate for various EC applications, where long-lasting performance and durability are key prerequisites. These findings contribute to the growing body of knowledge in the field of EC materials, offering exciting possibilities for their practical implementation in next-generation technologies.

5. Conclusions

This study has significantly advanced our understanding by systematically investigating the influence of film thickness, annealing temperature, and pH on the EC properties of electrodeposited TiO_2 thin films. Utilizing simple electrodeposition method, we successfully fabricated TiO_2 films on ITO glass, catering to potential applications in smart windows. Our comprehensive examination of synthesis parameters has unveiled optimal conditions that yield exceptional EC performance. In particular, a TiO_2 film prepared in an acidic environment and annealed at 450°C demonstrated outstanding attributes. The structural and morphological analysis indicates the emergence of the anatase phase, characterized by island-like crack morphology. This structural configuration proves conducive to the efficient diffusion of ions and electrons, thereby enhancing the EC features of the fabricated TiO_2 films. The films exhibited a substantial ΔT of 17.18%, a high CE of 58.8 cm^2/C, and swift switching kinetics. These numerical results underscore the critical importance of precise control over film preparation parameters. The obtained data not only enhances the scientific understanding of TiO_2's EC behavior but also provide a solid foundation for future research and development in the field, emphasizing the significance of optimized synthesis conditions for achieving superior electrochromic performance in TiO_2 thin films.

Acknowledgements

This research was partially supported by The Institute of Science, Dr. Homi Bhabha State University, 15, Madam Cama Road, Mumbai-400032, MH, India.

Author contributions

Ayesha Khan performed the experimental work; Dr. Anamika V. Kadam reviewed, revised, and edited the manuscript. Ayesha Khan has analyzed the data and wrote the manuscript. All authors participated in refining the manuscript and gave their approval for the final submitted version.

Conflict of interest

The authors declare no conflict of interest.

Data availability

The datasets generated during and analyzed during the current study are available from the corresponding author upon reasonable request.

Author details

Ayesha Khan and Anamika Vitthal Kadam*
The Institute of Science, Dr Homi Bhabha State University, Mumbai, India

*Address all correspondence to: anamiak@iscm.ac.in

References

[1] Granqvist CG. Electrochromics for smart windows: Oxide-based thin films and devices. Thin Solid Films. 2014;**564**:1-38. DOI: 10.1016/j.tsf.2014.02.002

[2] Ohsuku T, Hirai T. An electrochromic display dioxide based on titanium dioxide. Electrochimica Acta. 1982;**27**:1263-1266. DOI: 10.1016/0013-4686(82)80146-1

[3] Deb SK. Opportunities and challenges in science and technology of WO_3 for electrochromic and related applications. Solar Energy Materials & Solar Cells. 2008;**92**:245, 258. DOI: 10.1016/j.solmat.2007.01.026

[4] Qiu MJ, Sun P, Zhang B, Yu JH, Fu Y, Yu X, et al. Reliable information encryption and digital display applications based on multistate smart windows. Advanced Optical Materials. 2018;**6**:1800338

[5] Liptay W. Electrochromism and Solvatochromism. Angewandte Chemie International Edition. 1969;**8**:177-187

[6] Wang JM, Sun XW, Jiao ZH. Application of nanostructures in EC materials and devices: Recent Progress. Materials. 2010;**3**:5029

[7] Xie Z, Gao L, Liang B, Wang X, Chen G, Liu Z, et al. Fast fabrication of a WO3•2H2O thin film with improved electrochromic properties. Journal of Materials Chemistry. 2012;**22**:19904

[8] Zhang J, Tu JP, Zhang D, Qiao YQ, Xia XH, Wang XL, et al. Multicolor electrochromic polyaniline-WO3 hybrid thin films: One-pot molecular assembling synthesis. Journal of Materials Chemistry. 2011;**21**:17316

[9] Neil S, Lou P, Hong Y, Brad P. Electrochromic dynamic windows for office buildings. International Journal of Sustainable Built Environment. 2012;**1**:125-139. DOI: 10.1016/j.ijsbe.2012.09.001

[10] Zuo R, Gaoxiang D, Zhang W, Liu L, Liu Y, Mei L, et al. Photocatalytic degradation of methylene blue using TiO2 impregnated diatomite. Advances in Materials Science and Engineering. 2014;**2014**:1-7. DOI: 10.1155/2014/170148

[11] Ghicov A, Albu SP, Macak JM, Schmuki P. High-contrast electrochromic switching using transparent lift-off layers of self-organized TiO2 nanotubes. Small. 2008;**4**:1063-1066

[12] Hosaka N, Sekiya T, Satoko C, Kurita S. Optical properties of single-crystal Anatase TiO_2. Journal of Physical Society of Japan. 1997;**66**:877-880

[13] Cronemeyer DC. Electrical and optical properties of rutile single crystals. Physics Review. 1952;**87**:876

[14] Prasai B, Bin Cai M, Underwood K, Lewis JP, Drabold DA. Properties of amorphous and crystalline titanium dioxide from first principle. Journal of Materials Science. 2012;**47**:7515-7752

[15] Kaur K, Singh CV. Amorphous TiO2 as a photocatalyst for hydrogen production: A DFT study of structural and electronic properties. Energy Procedia. 2012;**29**:291-299. DOI: 10.1016/j.egypro.2012.09.035

[16] Zou JA, Gao JC, Xie FY. An amorphous TiO_2 sol sensitized with H_2O_2 with the enhancement of photocatalytic activity. Journal of Alloys

and Compounds. 2010;**497**:420-427. DOI: 10.1016/j.jallcom.2010.03.093

[17] Randorn C, Irvine JTS, Robertson P. Synthesis of visible-light-activated yellow amorphous TiO2 Photocatalyst. International Journal of Photoenergy. 2008;**2008**:1-6. DOI: 10.1155/2008/426872

[18] Kanna M, Wongnawa S, Buddee S, Dilokkhunakul K, Pinpithak P. Amorphous titanium dioxide: A recyclable dye remover for water treatment. Journal of Sol-Gel Science and Technology. 2010;**53**:162-170. DOI: 10.1007/s10971-009-2072-5

[19] Jeong HY, Lee JY, Choi SY. Interface-engineered amorphous TiO_2-based resistive memory devices. Advanced Functional Materials. 2010;**20**:3912-3917. DOI: 10.1002/adfm.201001254

[20] Chen X, Mao SS. Titanium dioxide nanomaterials: Synthesis, properties, modifications and applications. Chemical Reviews. 2007;**107**(7):2891-2959. DOI: 10.1021/cr0500535

[21] Li W, Ni C, Lin H, et al. Size dependence of thermal stability of TiO2 nanoparticles. Journal of Applied Physics. 2004;**96**:6663-6668. DOI: 10.1063/1.1807520

[22] Ding K, Miao Z, Hu B, An G, Sun Z, Han B, et al. Study on the anatase to rutile phase transformation and controlled synthesis of rutile nanocrystals with the assistance of ionic liquid. Langmuir. 2010;**26**(12):10294-10302. DOI: 10.1021/la100468e

[23] Shen L, Bao N, Zheng Y, et al. Hydrothermal splitting of titanate fibers to single-crystalline TiO2 nanostructures with controllable crystalline phase, morphology, microstructure, and photocatalytic activity. Journal of Physical Chemistry C. 2008;**112**:5985-5994. DOI: 10.1021/jp711369e

[24] Petkov V, Holzhüter G, Tröge U, et al. Atomic-scale structure of amorphous TiO2 by electron, X-ray diffraction and reverse Monte Carlo simulations. Journal of Non-Crystalline Solids. 1998;**231**:17-30. DOI: 10.1016/S0022-3093(98)00418-9

[25] Ibrahim SA, Sreekantan S. Effect of pH on TiO_2 nanoparticles via sol-gel method. Advanced Materials Research. 2011;**173**:184-189

[26] Hakki HK, Allahyari S, Rahemi N, Tasbihi M. The role of thermal annealing in controlling morphology, crystal structure and adherence of dip coated TiO2 film on glass and its photocatalytic activity. Materials Science in Semiconductor Processing. 2018;**85**:24-32. DOI: 10.1016/j.mssp.2018.05.031

[27] Kim C, Lee S, Kim S, Yoon J. Effect of annealing temperature on the capacitive and oxidant-generating properties of an electrochemically reduced TiO2 nanotube array. Electrochimica Acta. 2016;**222**:1578-1584. DOI: 10.1016/j.electacta.2016.11.143

[28] Lin SY, Chen YC, Wang CM, Liu CC. Effect of heat treatment on electrochromic properties of TiO 2 thin films. Journal of Solid-State Electrochemistry. 2008;**12**:1481-1486. DOI: 10.1007/s10008-007-0502-2

[29] Verma A, Basu A, Bakhshi AK, Agnihotry SA. Structural, optical and electrochemical properties of sol-gel derived TiO2 films: Annealing effects. Solid State Ionics. 2005;**176**:2285-2295. DOI: 10.1016/j.ssi.2005.06.011

[30] Sadiq SA, Waseem Z, Hanif S, Riaz S, Hayat K, Batool AIF, et al. Investigation

of the role of pH on structural and morphological properties of titanium dioxide nanoparticles. In: IOP Conference Series: Materials Science and Engineering. Vol. 863. Institute of Physics Publishing; 2020. p. 012046. DOI: 10.1088/1757-899X/863/1/012046

[31] Xia L, Feng ZH, Dai HT, et al. The influence of pH of the precursor solution on TiO2 films under hydrothermal synthesis, advances in display technologies III. SPIE. 2013;**8643**:86430L. DOI: 10.1117/12.2003018

[32] Jahromi HS, Taghdisian H, Afshar S, Tasharrofi S. Effects of pH and polyethylene glycol on surface morphology of TiO2 thin film. Surface and Coating Technology. 2009;**203**:1991-1996. DOI: 10.1016/j.surfcoat.2009.01.034

[33] Dinh NN, Oanh NTT, Long PD, et al. Electrochromic properties of TiO2 anatase thin films prepared by a dipping sol-gel method. Thin Solid Films. 2003;**423**:70-76. DOI: 10.1016/S0040-6090(02)00948-3

[34] Nang Dinh N, Minh Quyen N, Chung DN, et al. Highly-efficient electrochromic performance of nanostructured TiO2 films made by doctor blade technique. Solar Energy Materials and Solar Cells. 2011;**95**:618-623. DOI: 10.1016/j.solmat.2010.09.028

[35] Patil RA, Devan RS, Liou Y, Ma YR. Efficient electrochromic smart windows of one-dimensional pure brookite TiO2 nanoneedles. Solar Energy Materials and Solar Cells. 2016;**147**:240-245. DOI: 10.1016/j.solmat.2015.12.024

[36] Chen JZ, Ko WY, Yen YC, et al. Hydrothermally processed TiO2 nanowire electrodes with antireflective and electrochromic properties. ACS Nano. 2012;**6**:6633-6639. DOI: 10.1021/nn300787r

[37] Ye T, Xiang Y, Ji H, et al. Electrodeposition-based electrochromic devices with reversible three-state optical transformation by using titanium dioxide nanoparticle modified FTO electrode. RSC Advances. 2016;**6**:30769-30775. DOI: 10.1039/c6ra03315f

[38] Akkurt N, Pat S, Mohammadigharehbagh R, Özgür M, Demirkol U, Olkun A, et al. Investigation of TiO2 thin films as a cathodic material for electrochromic display devices. Journal of Materials Science: Materials in Electronics. 2020;**31**:9568-9578. DOI: 10.1007/s10854-020-03499-0

[39] Nunes D, Freire T, Barranger A, Vieira J, Matias M, Pereira S, et al. TiO2 nanostructured films for electrochromic paper based-devices. Applied Sciences. 2020;**10**:1200(1-14). DOI: 10.3390/app10041200

[40] Eyovge C, Deenen CS, Ruiz-Zepeda F, Bartling S, Smirnov Y, Morales-Masis M, et al. Color tuning of electrochromic TiO2Nanofibrous layers loaded with metal and metal oxide nanoparticles for smart Colored windows. ACS Applied Nano Materials. 2021;**4**:8600-8610. DOI: 10.1021/acsanm.1c02231

[41] Almarri AH. Enhanced electrochromic properties of anatase TiO_2 for flexible electrochromic device. Ionics. 2022;**28**:4435-4444. DOI: 10.1007/s11581-022-04646-9

[42] Xing C, Yang L, He R, Spadaro MC, Zhang Y, Arbiol J, et al. Brookite TiO2 nanorods as promising electrochromic and energy storage materials for smart windows. Small. 2023;**19**(49):2303639(1-11). DOI: 10.1002/smll.202303639

[43] Aiempanakit C, Junbang P, Suphap W, Aiempanakit K. Influence of water content on structural and electrochromic

properties of TiO2 nanotube prepared by anodization, journal of metals. Materials and Minerals. 2023;**33**:1680. DOI: 10.55713/jmmm.v33i3.1680

[44] Auer A, Steiner D, Portenkirchner E, Kunze-Liebhäuser J. Non-equilibrium phase transitions in amorphous and anatase TiO2 nanotubes. ACS Applied Energy Materials. 2018;**1**(5):1924-1929. DOI: 10.1021/acsaem.7b00319

[45] Fu X, Jia C, Wan Z, Weng X, Xie J, Deng L. Hybrid electrochromic film based on polyaniline and TiO2 nanorods array. Organic Electronics. 2014;**15**:2702-2709

[46] Ma Y, Ni M, Li S. Optimization of malachite green removal from water by TiO2 nanoparticles under UV irradiation. Nanomaterials. 2018;**8**:428(1-11). DOI: 10.3390/nano8060428

[47] Chaopeng F, Foo C. Pooi see Lee, one-step facile electrochemical preparation of WO_3/graphene nanocomposites with improved electrochromic properties. Electrochimica Acta. 2014;**117**:139-144

[48] Puchert MK, Timbrell PY, Lamb RN. Postdeposition annealing of radio frequency magnetron sputtered ZnO films. Journal of Vacuum Science & Technology A: Vacuum, Surfaces, and Films. 1996;**14**:2220-2230. DOI: 10.1116/1.580050

[49] Mahe M, Heintz JM, Rödel J, Reynders P. Cracking of titania nanocrystalline coatings. Journal of the European Ceramic Society. 2008;**28**:2003-2010. DOI: 10.1016/j.jeurceramsoc.2008.02.002

[50] Leftheriotis G, Yianoulis P. Development of electrodeposited WO3 films with modified surface morphology and improved electrochromic properties. Solid State Ionics. 2008;**179**:2192-2197. DOI: 10.1016/j.ssi.2008.07.018

[51] Mandlekar BK, Jadhav AL, Jadhav SL, Khan A, Kadam AV. Binder-free room-temperature synthesis of amorphous-WO3 thin film on FTO, ITO, and stainless steel by electrodeposition for electrochromic application. Optical Materials. 2023;**136**:113460. DOI: 10.1016/j.optmat.2023.113460

[52] Wen X, Zhang H. Photoelectrochemical properties of CuSGeO2-TiO2 composite coating electrode. PLoS One. 2016;**11**:e0152862. DOI: 10.1371/journal.pone.0152862

[53] Saravanan L, Kumar RM, Arumugum P, Jayavel R. Synthesis and photophysical studies of PVP capped Titania Nanostrips for photocatalytic applications. Journal of Optoelectronics and advanced materials. 2010;**4**:1676-1680

[54] Zhang JY, Boyd IW, O'Sullivan BJ, et al. Nanocrystalline TiO2 films studied by optical, XRD and FTIR spectroscopy. Journal of Non-Crystalline Solids. 2002;**303**:134-138. DOI: 10.1016/S0022-3093(02)00973-0

[55] Qiu J, Li S, Gray E, Liu H, Gu QF, Sun C, et al. Hydrogenation synthesis of blue TiO2 for high-performance lithium-ion batteries. Journal of Physical Chemistry C. 2014;**118**:8824-8830. DOI: 10.1021/jp501819p

[56] Babu SJ, Rao VN, Murthy DHK, Shastri M, Murthy M, Shetty M, et al. Significantly enhanced cocatalyst free H2 evolution from defect-engineered Brown TiO2. Ceramics International. 2020;**47**(10):14821-14828. DOI: 10.1016/j.ceramint.2020.10.026

[57] Dhandayuthapani T, Sivakumar R, Ilangovan R, Gopalakrishnan C,

Sanjeeviraja C, Sivanantharaja A, et al. Efficient electrochromic performance of anatase TiO2 thin films prepared by nebulized spray deposition method. Journal of Solid-State Electrochemistry. 2018;**22**:1825-1838

[58] Usharani S, Rajendran V. Optical, magnetic properties and visible light photocatalytic activity of CeO2/SnO2 nanocomposites. Engineering Science and Technology, an International Journal. 2016;**19**:2088-2093. DOI: 10.1016/j.jestch.2016.10.008

[59] Bakiro M, Ahmed SH, Alzamly A. Effect of pH, surfactant, and temperature on mixed-phase structure and band gap properties of BiNbO4 nanoparticles prepared using different routes. Chemistry (Switzerland). 2019;**1**:89-110. DOI: 10.3390/chemistry1010008

[60] Ozer N. Reproducibility of the coloration processes in TiO_2 films. Thin Solid Films. 1992;**214**(1):l-24

[61] Moulder JF. Handbook of X-Ray Photoelectron Spectroscopy. Physical Electronics Division: Perkin-Elmer Corporation; 1992

[62] Sivakumar R, Raj AME, Subramanian B, et al. Preparation and characterization of spray deposited n-type WO3 thin films for electrochromic devices. Materials Research Bulletin. 2004;**39**:1479-1489. DOI: 10.1016/j.materresbull.2004.04.023

Section 2

Medical Applications

Chapter 5

A Brief Overview and Application of Nickel-Titanium Shape Memory Alloy in Dentistry

Minja Miličić Lazić, Marijana Popović Bajić, Igor Đorđević, Marija Živković, Vojkan Lazić, Vukoman Jokanović, Ilija Nasov and Slavoljub Živković

Abstract

Nickel-titanium alloys are the most common shape memory alloys (SMA) widely used in dentistry. The attractive properties of Ni-Ti BioSMAs (shape memory effect, superelasticity, good corrosion, wear resistance, and good biocompatibility) make them highly desirable candidates for the design of implants, prosthetic components, orthodontic wires, and endodontic files. The aim of this chapter is to present the advantages of Ni-Ti alloy in dentistry through a selection of optimal chemical composition and various surface treatments (mechanical polishing, electrochemical polishing, chemical etching in acid solutions, heat treatment). The osteoconductivity of the Ni-Ti alloy on human bone cells and the anticorrosive and antibacterial effects of nanocoating orthodontic wires were tested. The results on human bone cells suggested good adhesion and proliferation of osteoblast-like cells to nickel-titanium surfaces, which could improve osseointegration of the potential dental implants. A study of magnetron sputtering method (MS) with TiN-Cu-nanocoatings on orthodontic archwires showed favorable results in corrosion resistance and antibacterial properties. Future improvement of mechanical properties is focused on applying various methods of surface modifications, especially applying coatings with antibacterial properties.

Keywords: nickel-titanium alloy, dental implants, endodontic files, orthodontic wire, corrosion properties, antibacterial properties

1. Introduction

Nickel-titanium is the only SMA with valuable biomedical applications among the different SMAs. Properties of nitinol that suit its application in dentistry are good biocompatibility, corrosion resistance, superelasticity, and fatigue resistance. This biomaterial is widely used in all fields of dentistry: prosthetics, orthodontics, and endodontics.

Even though Ni-Ti medical devices were first introduced in the early 70s, it passed 20 years since that discovery and wider commercial use was established [1]. During the 80s, among various experimental efforts, the scientific community was full of skepticism about the biological safety of these alloys, considering the fact that equiatomic nickel-titanium alloys consist of around 50 weight percent of nickel. At that time, nickel sensitivity, toxicity, and carcinogenicity were widely discussed [2].

With further understanding and improvement of the technological production processes and the corrosion resistance of these alloys, a more comprehensive commercial application began. Yet, many studies conducted during the mid-1990s indicated good biocompatibility, thanks to encouraging findings of corrosion resistance of nitinol equiatomic or near equiatomic alloy. The same decade was known for the advent of nitinol stent [3, 4]. It was reported that formation of the protective surface layer could prevent nickel ion release [5]. The nickel-free zone within the upper surface and the presence of homogenous titanium oxide are the main factors defining good biocompatibility of this alloy.

Besides inevitable biological safety, any material considered for biomedical use must possess biofunctional properties, also known as biomechanical compatibility with the surrounding environment in the human body. Two unique functional properties, pseudoelasticity and shape memory effect, are considered to be responsible for this alloy's biomechanical compatibility. Also, favorable properties are compressive strength (higher than human bone) and elastic modulus (almost similar to osseous tissue) [6].

Due to thermoelastic martensitic transformation, this alloy can adapt to the conditions of the environment. Crystal state defines the physical properties of the material. The austenitic form is rigid, contrary to the ductile martensitic, which is recognized as a distinctive feature, i.e., the alloy's capability to be adjusted to the desired application. Phase transformations are characterized by diffusion-less, coordinated movement of the atoms [7]. It occurs while receiving or releasing minimum energy for a given temperature or mechanical load. This property of the Ni-Ti alloy enables its applications in orthodontics and endodontics.

Bearing in mind that the functional properties of these alloys are sensitive and highly dependent on chemical composition changes, the fabrication of this device for biomedical use must fulfill requirements based on Ni/Ti ratio. Consequently, by changing the alloy composition to 0.1%, the transformation temperature can be changed to 10°C [8]. More precisely, higher Ti content increases the transformation temperature. Oppositely, increasing the Ni content by 1%, *Ms* temperature decreases significantly, at a rate of about 100°C [9]. Also, these minor changes in composition make considerable changes in performance characteristics. Titanium-rich alloys are known for their excellent shape recovery (i.e., thermal phase transformation phenomenon).

In contrast, nickel-rich Ni-Ti alloys gained attention for their superelastic behavior (i.e., stress-triggered phase transformation), which is why most Ni-Ti biomedical applications rely on this feature. Moreover, higher nickel content has been reported to improve mechanical properties [10], and yield strength is the most important for biomedical purposes. The addition of heavy metals, like nickel, can reduce grain size during solidification. Besides improving mechanical properties, grain refining enhances surface characteristics and consequently improves tear resistance.

2. Features of Ni-Ti alloys

2.1 Shape memory effect

This unique property was discovered in the early 60s by Buehler and his co-workers at the U.S. Naval Ordnance Laboratory. This feature is recognized as the ability of a material to transform its macroscopic shape upon temperature changes. Depending on the reversibility, the shape memory phenomenon can exist as a one-way or two-way SME. If the alloy is permanently mechanically deformed at low temperatures (cold state), it returns to its original shape upon heating [11]. In this one-way phenomenon, the material can be deformed to its original shape upon heating. However, the two-way shape memory effect is characterized by reversible transformations from low to high temperatures. TWSME is not an ordinary material property, as alloy must undergo specific training to obtain this property [12].

In both cases, OWSME or TWSME alloy starts to deform below the martensite start temperature, *Ms*.

2.2 Superelasticity (pseudoelasticity)

The pseudoelastic feature of nitinol was discovered in the early 1970s and represented materials' properties to reach elastic deformation up to 8% [6]. Even though the term superelastic is commonly used in literature, the pseudoelastic phenomenon is more convenient when explaining this functional property. When the material is exposed to stress at temperatures above *Af*, nonlinear deformation without residual strain is developed.

Figure 1a represents a typical stress–strain diagram for conventional metallic materials, and it is visible that upon loading, stress increases linearly. Contrary to this conventional behavior, the stress-strain diagram for pseudoelastic Ni-Ti alloy is shown in **Figure 1b**. The interval from 0 to A represents behavior when stress increases

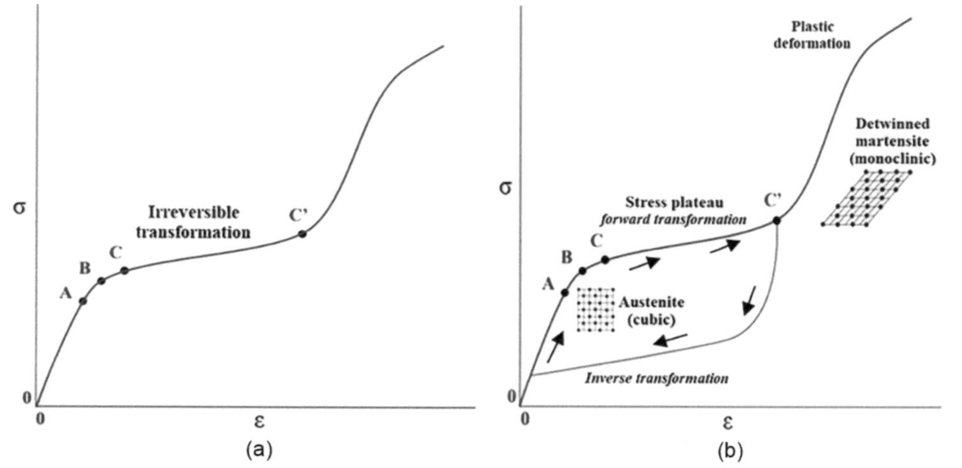

Figure 1.
Stress-strain diagram: (a) typical curve for conventional metallic materials, A-Linear elasticity, B-Elastic limit, C-Yield point, C'-Fracture point; (b) Ni-Ti alloy with inverse transformation behavior, that is, "flag" of a diagram, A-Austenite phase, B-Martensite twinned phase, C-C'-Non-linear deformation i.e. non-linear behavior of stress and strain, C'-Martensite detwinned phase.

linearly with a strain upon loading. Point B is the yield point. The interval from C to C′ is a loading plateau (i.e., stress plateau) where large strains are developed with a slight stress increase. At point C, the Ni-Ti alloy can reach an 8% strain. Concurrently, the structure lattice (point C′) transforms in stress-induced martensite (monoclinic crystal lattice), and the material possesses that crystallographic form until the stress is removed. After unloading, it reverts into parent austenite because the inverse transformation occurs. The strain is recovered at the lower point of the unloading plateau, and the material exhibits a body-centered cubic lattice (austenite phase). In conclusion, the uniquenesses of the pseudoelastic Ni-Ti alloy diagram is a hysteresis loop (also known as a stress-strain flag), and the size of this loop can vary regarding the variation in chemical composition or thermomechanical treatment of the material.

2.3 Corrosion resistance

In terms of a wide array of medical applications, the corrosion resistance of nitinol medical devices is a crucial factor in defining their biocompatibility. Corrosion of nickel-titanium alloys is a contextual phenomenon (**Figure 2**), depending on surface treatments and the nature of the environment [4]. The investigation of the influence of pH change on corrosion is well documented. These alloys are corrosion-resistant in pH-neutral conditions [13].

Nickel-titanium alloys are known for their passivization state. Still, in order to explain the overall mechanisms of corrosion resistance of nitinol alloy, the most important fact is that the human body presents very challenging surroundings. The reason for that is the presence of dissolved oxygen, chlorides, and changes in pH levels (acid-base disbalance). Moreover, different parts of the human body vary significantly in oxygen concentrations and pH values. Various studies have demonstrated lower corrosion resistance of these alloys in acidic solutions and chloride-containing environments [14–16]. In general, it is well documented that the corrosion potential increases as the pH value decreases [17]. Also, an increase in the environment temperature decreases surfaces' ability to repassivate.

On the contrary, Montero-Ocampo and his co-workers reported the beneficial effects of heat treatment (annealing) on corrosion resistance [6]. Results of their

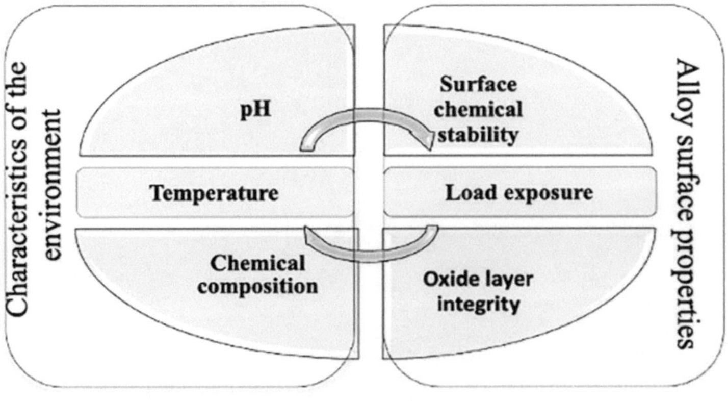

Figure 2.
Corrosion dependent variables.

study showed that microstructure evolution generated corrosion density hysteresis of almost zero during thermomechanical processing and a pitting potential of >1 [18].

According to the attack mechanism, Ni-Ti devices are more prone to localized forms of corrosion and pitting. When exposed to aggressive anionic species, local breakdown occurs [19, 20]. The localized surface area becomes anodic, while the rest of the alloy surface is cathodic (**Figure 3**). Further, pit propagation is driven by an autocatalytic process, which means that once the aggressive anion penetrates the surface, the surroundings become depleted in oxygen [21, 22]. Pits grow in the direction of gravity, producing an excess of positive charges inside the pit. These conditions initiate more migration of anions into the localized defect, causing its propagation.

Numerous studies point out chloride ions as one of the most aggressive agents responsible for the degradation of passive alloys, such as nickel-titanium, from the aspect of the individual effect of elements within corrosion compounds. The reason for this is, above all, the small size of this molecule that allows it to penetrate more easily into the surface oxide layer. In addition, the absorption of Cl^- ions is easy as it has a negative charge on the positively charged oxide layer [23, 24]. After the initial penetration of chloride into the surface layer, further outcomes can go in two directions, repassivation and complete or partial recovery, or further propagation.

The possibility of repassivation, that is, the ability of the oxide layer to perform self-healing after surface damage, is crucial for its long-term stability. This ability depends on the quality of the oxide film. The homogeneity of the oxide will ensure continuity and prevent metal leakage ions. Otherwise, the inhomogeneous oxide on the surface will allow the accumulation of hydrogen ions inside the initially formed crack, which creates a more positive potential value (**Figure 3**).

2.4 Surface chemical stability

The surface characteristics of Ni-Ti alloy are highly dependent on the preparation method. Literature-based evidence is coherent about the statement that the Ni/Ti ratio on the surface can vary significantly [25]. Also, during exposure to the corrosive environment, the thickness of the oxide on the surface of the nickel-titanium alloy can be in a wide range of nanometer scale, from 4 to 3500 nm [26]. Several studies have analyzed the oxide layer growth potential on the sample's surface. Results showed that untreated samples developed a thicker oxide on the surface (120–340 nm), while on mechanically polished or etched samples, the film was up to 10 times thinner (11–16 nm) [27]. However, the protective role of the surface oxide layer against corrosive agents is

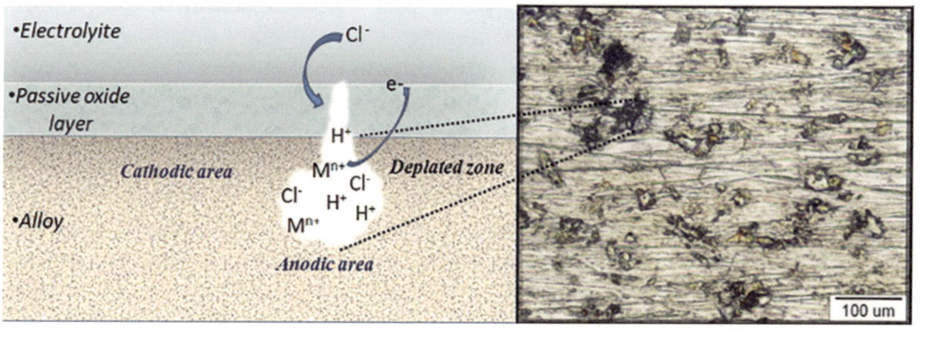

Figure 3.
Mechanism of pitting corrosion in Cl^--containing environment.

determined by the ability of the alloy to develop a homogeneous and compact oxide. The integrity of this layer is much more important than its thickness [28].

To date, various surface treatments have been investigated. The most common are traditional methods, which include mechanical polishing, electrochemical polishing, chemical etching in acid solutions, and heat treatment (treatment in an autoclave) [29]. In addition, there are plasma ion implantation methods and bioactive coatings formation [25].

Untreated samples develop a surface layer of oxygen, carbon, titanium oxide, and small amounts of nickel. Conventional preparation, such as mechanical polishing, can increase titanium concentrations up to 5 times [4].

Also, polished surfaces consist of a combination of titanium, titanium oxide, a small amount of nickel oxide, and elemental nickel. In contrast, the oxygen concentration is significantly lower on chemically etched surfaces. Previous studies [30, 31] indicated multiple advantages of polishing over chemical etching. It was concluded that etching with various acids can create pitting defects up to 8 μm deep. Such roughness can favor local corrosion.

Besides, if the alloy is autoclaved in water, the titanium ratio increases 20 to 30 times more than nickel. As the immersion duration increases, more Ti develops on the surface. Consequently, nickel concentrations decrease [4].

The process of spontaneous formation of transition metal oxides is measured by the Gibbs free energy [32]. The more significant presence of Ti oxide on the surface of nickel-titanium alloys is explained by its formation requiring four times less energy than Ni oxide.

Given that *in vivo* implantation is accompanied by severe environmental factors (body fluids- blood, saliva, acid-base imbalance, body salts, etc.), the surface structure may differ significantly from that *in vitro*. In line with that, Hanawa and his co-workers [33] showed that nitinol devices implanted in human bone develop a thin calcium phosphate (Ca/P) layer in combination with titanium oxide. Results of further investigations confirmed that this layer is considered responsible for nitinol's good biocompatibility.

3. Application of Ni-Ti alloy in dentistry

3.1 Application of Ni-Ti alloy in prosthodontics

To join the advantages and overcome the limitations of the screw and cement-retained prosthetic components, a new retention system based on the two-way shape memory effect (TWSME) has been developed lately, etc. nitinol sleeve attachment component.

To be able to solve the problem of inadequate dental implant position and to achieve adequate angulation of the superstructure in implant-prosthodontics, nickel-titanium was used for the first time to manufacture a memory abutment in the 90s of the last century (Dyna memory abutment) [34]. But, this abutment was compatible only with Dyna implants. Due to limited application and high production costs, this system did not achieve wide commercial use.

Contemporary product design (Rodo abutment, Smileloc® Abutment System, Nitinol sleeve) and machine-precise components lead to widespread application. This component is compatible with different implant systems (Straumann, Neodent, Nobel Biocare, Biomet, and Biohorizons). The transition to the Rodo abutment (Smileloc® Abutment System) led to the appearance of a slightly more advanced abutment system with a memory shape (**Figure 4**). As part of a crown on implants,

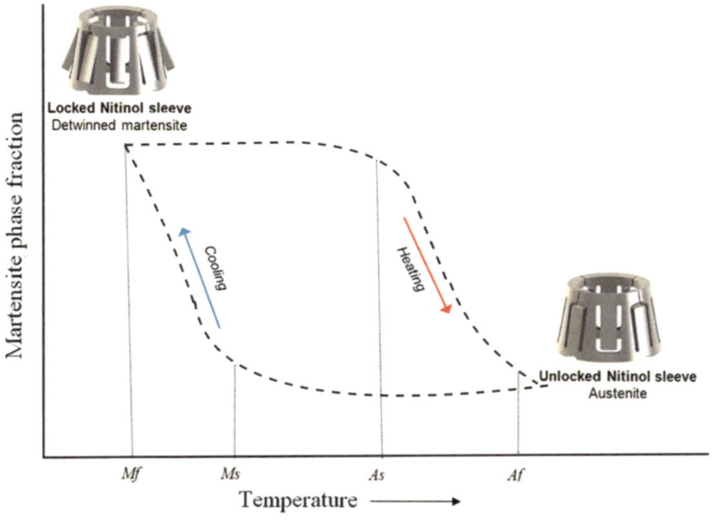

Figure 4.
Two-way shape memory mechanism of nitinol abutment sleeve.

a nitinol sleeve over the abutment has recently been used as an alternative to screw-retained or cement-retained crowns to overcome technical and biological complications in implant prosthetics. This system, familiar to clinicians for the last 10 years, is characterized by a retention nitinol component placed between the implant superstructure and the crown. The nitinol component is the precision-machined sleeve [35]. The retention component can persist in two forms, i.e., austenite cubic structure (at high temperatures) and martensitic body-centered tetragonal crystal structure (at low temperatures). **Figure 4** represents the working principle of the shape memory alloy sleeve. The temperature change leads to changes within the crystal lattice resulting in a macroscopic change in the shape of the component from an "unlocked" to a "locked" form.

According to the data from X-ray images taken after 6 months of follow-up, crowns retained on the nitinol sleeve were kept in the same position without any restoration movement [36]. Although these results suggested stable retention, evidence of long-term clinical studies is needed to evaluate optimal retention force, especially in pro arch implant prosthodontics restorations.

When martensite form is heated, it begins to change into austenite (**Figure 4**). The temperature at which this phenomenon starts is called austenite start temperature (*As*). The temperature at which this phenomenon is complete is called austenite finish temperature (*Af*). When the austenite form of the sleeve is cooled, it begins to change into martensite. The temperature at which this phenomenon starts is called martensite start temperature (*Ms*). The temperature at which martensite is again completely reverted is called martensite finish temperature (*Mf*). Hysteresis is generally defined as the difference between the temperatures at which the material is 50% transformed to austenite upon heating and 50% transformed to martensite upon cooling. This difference can be up to 20–30°C [37].

The uniqueness of nickel-titanium alloy was used to develop a second type of abutment, the EZ Crown system. Even though this abutment consists of a nitinol component, its retention is based on features different from the shape memory sleeve abutment.

Namely, the superstructure of the EZ Crown internal component (attachment system) consists of a nitinol retention spring, the zirconia ball, and the cylindrical suprastructure (**Figure 5**) [39].

In this complex configuration, the nitinol spring is in a martensitic stabilized configuration, meaning that the product will not develop either pseudoelasticity or the shape memory effect. As seen from **Figure 5**, Zr balls are positioned into the undercut region of the abutment retention groove.

Once the Ni-Ti is placed over the Zr ball, it develops a stable retention force.

Thanks to the low modulus of elasticity and good springback capacity of the martensitic stabilized nickel-titanium component, the nitinol spring will provide a continuous force on the zirconia ball [40].

The first dental Ni-Ti implants were introduced in the 80s of the last century [41].

Important properties of the metallic biomaterials are that they exhibit similar elastic deformation behavior as local tissue in human environment. It is known that elastic deformation of Co-Cr alloys and stainless steel is limited to 1%. Contrary, human osseous tissue can be elastically deformed up to 10% strein. Nitinol is only commercially available metallic material that can behave in a similar way when exposed to mechanical deformation.

Regarding the desired properties of the materials for implant manufacture, preference should be given to alloys with a high damping capacity and a low modulus of elasticity. High damping capacity is the desired property when it comes to orthopedic materials. This feature represents materials' ability to absorb and evenly transfer mechanical forces, eliminating sudden shocks and oscillations.

Thanks to good flexibility, nitinol alloy can convert accepted mechanical energy into thermal energy, making the implant resistant to shocks caused by external loading. Therefore, metallic materials with high damping capacity values will result in a well-balanced stress-strain ratio within the alloy [42].

Regarding the modulus of elasticity, the desired property of the alloys is that these values should be as low as possible and close to those of Young's modulus of bone. The Ti6Al4V titanium alloy used for making implants has a modulus of elasticity of 110 GPa, and for pure titanium, that value is 116 GPa. The Co-Cr alloy has a modulus of elasticity

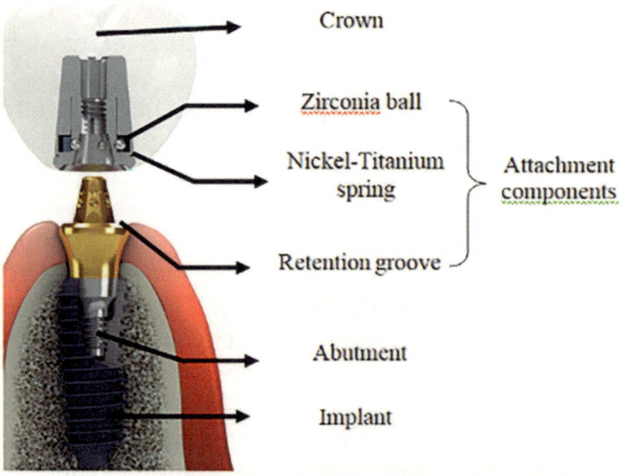

Figure 5.
Retention system of EZ crown [38].

of 210–235 GPa, and the nickel-titanium is 80 GPa for the austenitic and 30 GPa for the martensitic form of the alloy. Young's modulus for human bone is 20 GPa [43].

If the material's physical properties for making implants are not aligned with the purpose of these devices, it can lead to technical and biological complications. The low damping capacity of the alloy more often leads to problems with the implant itself (fracture), and differences in the values of the modulus of elasticity of the material for making the implant and the surrounding bone can develop unfavorable effects on the supporting tissue and decrease bone density [38].

Regarding the mechanical properties of Ni-Ti, this material is biomechanically compatible with the surrounding environment. In comparison to Ti-base implants, it was reported that Ni-Ti alloy exhibits higher biomechanical compatibility with human bone under extension flexion in the sagittal plane [11].

Lately, in bioengineering science, considerable efforts have been made to develop devices that mimic the natural human environment. Additive manufacturing enables the production of devices with desired porosity. This design is suitable for the production of scaffolds as a porous matrix allows for osteoblasts to grow.

However, data in the literature indicate the importance of precise defining the size and distribution of these pores because, in the case of the presence of irregular porosity of the material, the strength could be because, with the increase in pores, the sensitivity of the metal structure to cracks propagation also increase. The suggested application requires far more preclinical and clinical investigation of the osteoconductivity of Ni-Ti.

3.1.1 Experimental procedure

The aim was to evaluate osteoblast-like cell adhesion and proliferation in the presence of Ni-Ti alloy.

The study was carried out in accordance with the Declaration of Helsinki and approved by the Institutional Ethics Committee (School of Dental Medicine, University of Belgrade, approval no. 36/7).

Ni-Ti alloy was produced by a standard casting process, i.e. vacuum remelting. To be able to obtain sample dimensions for experimental testing, electro-erosion cutting was performed. The final experimental sample dimensions were 2r = 11 mm, thickness 1.7 mm, n = 5. Round-over glass with the same dimensions as Ni-Ti samples served as control. All of the samples were cleaned with acetone, alcohol, and deionized water for 90 seconds, blow-dried, sterilized with UV-C light for 1 hour, and transferred to a culture flask with 12 wells.

3.1.1.1 Cell culture

Osteoblast-like cells ([MG-63] ATCC® CRL-1427™, USA) were incubated under strictly controlled conditions (at 37°C in a humidified 5% CO_2 atmosphere) in a complete growth medium (Dulbecco's modified Eagle's medium (DMEM) with 4 mM L-glutamine, 10% Foetal bovine serum (FBS), and 1% ABAM, Sigma-Aldrich, Steinheim, Germany), and passaged every 3 days.

3.1.1.2 Adhesion pattern and cell viability

For the assessment of biocompatibility after direct exposure, osteoblast-like cells were seeded into each well at a density of 2 × 104 cells cm^{-2}. Human bone cells were plated on Ni-Ti samples and control cover glass. The incubation period was 24 h.

The number and the viability of attached osteoblast-like cells were evaluated using differential staining. In order to remove partially attached cells, the samples were washed with Phosphate-Buffered Saline. Then, samples were stained with 2 μg/mL Hoechst 33342 dye for 20 min, which bound to the DNA of live cells. The number of viable cells was assessed by emission of blue fluorescence and counted using ImageJ (NIH).

3.1.1.3 Preparation for scanning electron microscopy

After incubation, samples with attached cells were fixed, contrasted and submerged into 2.5% glutaraldehyde (SPI Supplies, West Chester, PA, USA) and 0.4% paraformaldehyde (Merck KGaA, Darmstadt, Germany) in 1 M Na-phosphate buffer (NaH2PO4·2H2O and Na2HPO4·2H2O; Merck KGaA, Darmstadt, Germany). Samples were gold-dusted using a Precision Etching Coating System (682 PECS, Gatan, Pleasanton, USA). The morphology and the attachment pattern of osteoblast-like cells were examined using a scanning electron microscope (JOEL JSM-6500F).

3.1.2 Results

3.1.2.1 Cell attachment and proliferation in the presence of Ni-Ti

The cell cultures showed that the cells had grown very close to the Ni-Ti surfaces (**Figure 6**). The shape of the attached cells was predominately elongated, with clearly visible large cell extensions suggesting good adhesion. Cells with membrane outgrowths predominate (pseudopodia and philopodia).

The results of fluorescence microscopy showed that after 24 h of exposure, tested Ni-Ti samples used were not cytotoxic, as the density of attached cells was almost comparable to the control indicating normal cell growth, and the results of staining cells with propidium iodide showed that the viability of attached cells in all samples was higher than 95% (98.6 ± 0.5% in control; 96.6 ± 0, 1% in Ni-Ti (**Figure 7**).

This study conducted on human bone cells suggested good susceptibility of osteoblast-like cells to nickel-titanium surfaces.

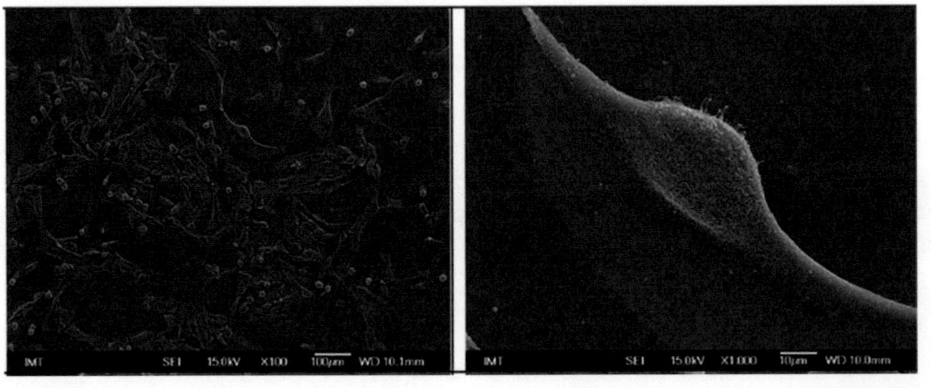

Figure 6.
Scanning electron microscopy of osteoblast-like cells cultured on Ni-Ti after 24 h.

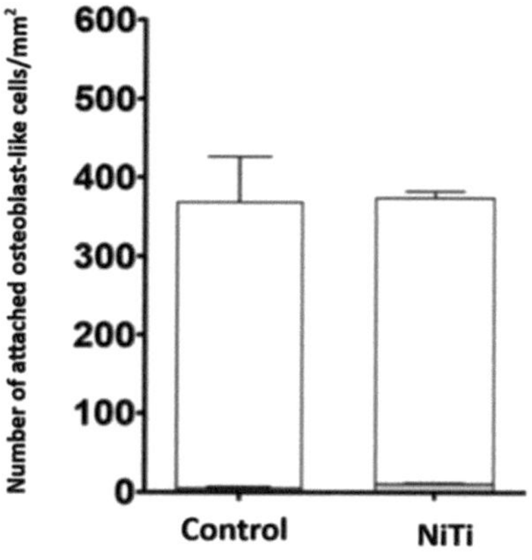

Figure 7.
Mean number of viable attached cells after 24 h.

3.2 Application of Ni-Ti alloy in orthodontics

In dentistry, the Ni-Ti alloy has the widest application for the production of orthodontic wires. In modern orthodontics, the choice of materials plays a significant role in achieving optimal outcomes. Among these materials, Ni-Ti archwires have gained prominence due to their unique mechanical properties.

Since they were originally introduced, the development of orthodontic archwires has continued, enhancing their properties in a range of clinical settings. The orthodontic archwires exert gentle and continual forces on teeth, which leads to tooth repositioning. When a force is applied over several weeks to months, the orthodontic archwire elastic properties of the archwire are needed. Furthermore, distinct orthodontic archwires are required for different stages of the orthodontic treatment. Due to their unique properties—the shape memory effect and superelasticity, Ni-Ti archwires can effectively correct the misalignment of teeth, which is extremely important in the initial phase of orthodontic treatment (leveling and alignment). When the force is transferred to the bone, the bone responds with resorption and apposition, allowing the tooth movement (**Figure 8**) [44].

Due to superelasticity, on unloading, Ni-Ti archwires may return to their original shape before loading. It is possible for the alloy to be deformed until 7–8% strain, which is almost forty times the capacity of the stainless steel wire. The loading leads to periodontal ligament deformation, and mobilization begins at a plateau of light forces recognized to provide the best biological tooth movement while the crystalline structure of the Ni-Ti alloy reverts to its initial configuration [45].

Since Ni-Ti alloy is a material with a shape memory effect, it has the property to "remember" its original shape after being elastically or pseudoplastically deformed by increasing its temperature. This effect results from thermoelastic martensitic transformation, where a body-centered cubic phase (austenitic phase) transforms into an orthorhombic or monoclinic martensite phase [46]. Therefore, in orthodontics, when the orthodontic archwire is deformed, it aligns and moves the teeth by returning to their natural shape. The stress remains almost at a constant level during the range of archwire activation, generating a low level of force, producing a good response from surrounding tissues, more comfort for the patient, and more physiological tooth movement. High orthodontic forces can produce root resorption, which is undesirable because it can affect the tooth's long-term viability [47]. Apart from using Ni-Ti

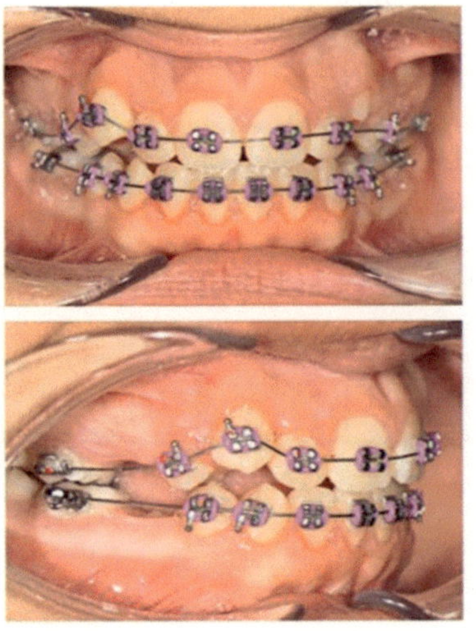

Figure 8.
The initial phase of orthodontic treatment (leveling and alignment) with Ni-Ti 0.012 inch archwire. Due to its superelastic properties, this archwire can tolerate a great degree of teeth misalignment.

alloys in archwires, they can be used in open coil springs, as well as closing loops. NiTi retraction loops for canine distalization are not widely used in everyday practice since their forming is not as easy as with other alloys. Open coil springs are used most often for molar distalization, space opening, or tooth uprighting, delivering light and continuous forces. However, these forces are very often below the optimal range, so the authors advise more activation of the Ni-Ti coils to take greater advantage of their superelastic properties [48]. The effectiveness of space closure with Ni-Ti coils compared to elastomeric chains is likely to be similar in clinical investigation [49]. However, the moderate quality evidence of a metanalysis suggests a faster space closure with Ni-Ti closing springs compared to elastomeric chains [50]. Force degradation of both elastomeric chains and Ni-Ti coils was noticed and should be considered by clinicians when choosing the manner of space closure [50].

Fixed appliances consist of orthodontic brackets, which are bonded to labial or lingual and orthodontic archwires, placed into the slots of orthodontic brackets to exert a force on the teeth. Therefore, oral hygiene is compromised due to greater adhesion of plaque and bacteria, and the risk of enamel decay and gingivitis is increased during the orthodontic treatment [51, 52]. Since silver nanoparticles are known for their antibacterial properties in numerous studies [53, 54], it is possible to design a bactericide archwire to help with dental plaque control. The electrodeposition of silver nanoparticles onto Ni-Ti orthodontic archwires can significantly reduce oral bacteria in bacterial culture by more than 90% without changing their calorimetric and mechanical properties and release of nickel; however, further clinical investigation of these findings is needed to prove these results [55].

The conventional Ni-Ti archwire shows a homogenous composition along its length. Functionally graded Ni-Ti shows varying composition along its length, allowing for tailored mechanical properties at different segments of the wire, which can provide customized force levels required for different teeth. The manufacturing of functionally graded Ni-Ti wires involves various methods. Microstructurally graded Ni-Ti can be created by variations in heat treatment conditions (annealing or aging temperature) along the length of the Ni-Ti wire or Ni-Ti plates across their thickness.

DOI: http://dx.doi.org/10.5772/intechopen.1004825

Compositionally graded Ni-Ti has a variation in composition (contents of Ni and Ti) within the body of the alloy. Also, the structure geometry can be purposely graded in the case of geometrically graded Ni-Ti archwires [56].

Also, two major problems in using NitTi alloys in orthodontics - friction and corrosion which lead to Ni-Ti release and mechanical weakening, can be encountered with different surface modifications of the archwire [57]. The results of a clinical study revealed that TiO_2 coating on the archwire surface decreased the surface roughness and adhesion of *Streptococcus mutans*. However, the advantage of coated archwire on surface roughness was lost after 1 month [58].

Future investigation should be focused on enhancing mechanical properties using different methods of surface modifications, preferably creating a coating with antibacterial properties.

3.2.1 Magnetron sputtering method

The Magnetron sputtering (MS) method is a state-of-the-art coating technology that relies on physical vapor deposition (PVD) in vacuum conditions [59]. MS enables rapid deposition of diverse metal/alloy coatings on the surface of various materials with improved adhesion, high purity of the obtained films and excellent coverage of the substrate features [60]. Overall, the use of this outstanding technology provides very high hardness, abrasion resistance, fine microstructure, as well as long-term chemical, thermal, and environmental stability, which is essential for materials sustainability and safe clinical use [59]. By applying MS, modification of the material's surface in single-layer, multilayer, or graded thin-layer coatings for obtaining systems with superior functional characteristics could be achieved.

In order to improve the properties of the Ni-Ti alloy in dentistry, this MS method was applied with protective coatings onto orthodontic archwires. The greatest challenge for safe orthodontic treatment is corrosion, chemically or microbiologically induced, resulting in the release of nickel (Ni). The aim of this investigation was to enhance resistance to corrosion and introduce antibacterial properties to Ni-Ti orthodontic archwires by coating them with copper (Cu) doper titanium nitride (TiN-Cu). Using cathodic arc evaporation (CAE), Ni-Ti archwires were coated with TiN-Cu and direct current magnetron sputtering (DC-MS). The morphology of the sample has been examined using field emission scanning electron microscopy (FESEM) whereas chemical composition was analyzed using energy-dispersive X-ray spectroscopy (EDS), X-ray diffraction (XRD) and Fourier transformed infrared spectroscopy (FTIR). To estimate the ion release, inductively coupled plasma optical emission spectrometry (ICP-OES) was used. The biocompatibility of samples was examined using 3-(4,5-dimethylthiazol-2-yl)-2,5-diphenyl tetrazolium bromide (MTT) assay. Antibacterial activity was tested for *Streptococcus mutans* and *Streptococcus mitis*.

The physicochemical characterization results showed that the coatings with the presence of TiN phase and incorporated Cu were well-designed. Such topography encourages corrosion resistance and consequently increases biocompatibility due to decreased surface roughness. EDS spectrum demonstrated the existence of Cu, Ti, and N in these samples, whereas XRD and FTIR analyses showed the presence of the TiN phase with incorporated Cu, indicating that the desired phases were obtained. TiN-Cu nanocoated archwires demonstrated a significantly reduced Ni release ($p < 0.05$). The 28-day eluates of TiN-Cu-nanocoated archwires had the highest relative cell viability ($p < 0.05$). Antibacterial tests showed a significant decrease in

Streptococcus mutans and *Streptococcus mitis* counts, regarding TiN-Cu-nanocoated archwires. In contrast, TiN coatings showed no reduction in *Streptococcus mutans*, indicating the importance of copper in the obtained results.

The most notable decrease in concentrations of Streptococcus mitis was found in the case of TiN-Cu-coated archwires ($p < 0.05$) (**Table 1**). Considering antibacterial tests and biocompatibility, TiN-Cu-nanocoated archwires may be a promising candidate for further clinical research [61].

3.3 Application of Ni-Ti alloy in endodontics

Thanks to the high flexibility of the alloy, nickel-titanium endodontic instruments represent the gold standard in root canal therapy. In order to ensure the possibility of simultaneous cleaning and shaping of complex morphology of the canal, the alloy must possess the property of superelasticity, and the characteristic unique only for binary alloys with a higher proportion of nickel (56 wt.%) compared to titanium (44 wt.%) [62]. Good flexibility and the possibility to be elastically deformed up to 8% are the main characteristics of endodontic instruments. Those properties ensure high fatigue resistance of endodontic files, thus preventing instrument breakage during the mechanical shaping of the root canal. Compared to stainless steel and cobalt alloys, good mechanical properties, favorable axial and torsional resistance, and lower values of the elastic modulus are all very important in the shaping of the root canal [14].

A big problem in endodontics is cleaning and shaping of the narrow and curved root canals. The intricate interplay of temperature, heating duration, and cooling rates during alloy production has been established to influence the superelasticity and shape memory of Ni-Ti files [63]. The application of heat treatment preserves the crystallographic structure of the alloy, imparting endodontic files with heightened flexibility and resistance to fractures [64].

Ni-Ti endodontic files predominantly exhibit either an austenitic phase (conventional Ni-Ti, M-wire, R-phase) or a martensitic phase (controlled memory (CM) wire, gold, and blue heat-treated Ni-Ti alloys) [45]. Specialized heat treatment has led to the development of superelastic alloys with a stable martensitic phase, showcasing a lower modulus of elasticity (30–40 GPa) compared to austenitic (80–90 GPa). Furthermore, the modulus in the R-phase is lower than in martensite [65]. The martensitic phase, with its double-phase structure reorientation, offers superior resistance to cyclic fatigue compared to austenitic phases [65, 66].

The production of conventional Ni-Ti wires involves a cold drawing process, yielding a microstructure that incorporates martensite residues within the austenitic matrix. To reduce internal stresses and drawbacks associated with the rearrangement of the crystal grid, a crucial step involves subjecting the alloy to heat treatment within

	Streptococcus mutans			*Streptococcus mitis*		
	TiN-Cu Nanocoated Archwires	**Ni-Ti Archwires**	**SS Archwires**	**TiN-Cu-Nanocoated Archwires**	**Ni-Ti Archwires**	**SS Archwires**
Log CFU ± SD	4.19 ± 0.22	4.29 ± 0.24	5.87 ± 1.10	3.08 ± 0.99	3.75 ± 0.82	5.54 ± 0.60

Table 1.
Final bacterial count for Streptococcus mutans *and* Streptococcus mitis.

the temperature range of 450–550°C [67]. Heat treatment enables endodontic files to be more resistant to breakage during root canal treatment.

Specific surface treatments play a pivotal role in enhancing the physical and mechanical properties of endodontic files. These treatments, such as electropolishing, electric discharge machining, ionic implementation, cryogenic treatment, and nitriding, contribute to the overall performance and longevity of endodontic files [14, 63].

Electropolishing is an electrochemical process for surface finishing Ni-Ti files. This method removes surface irregularities by dissolving metal ions in an electrolyte bath, creating a thin passive layer that enhances resistance to cyclic fatigue, torsional load, and corrosion [63–65]. RaCe systems and EndoSequence files are examples of electropolished files, offering improved cutting, reduced screwing inside the canal, and enhanced apical penetration [63].

In the ever-evolving landscape of endodontics, Electric Discharge Machining (EDM) technology has emerged as a transformative force, offering a non-contact thermal erosion process to produce electrically conductive materials. This cutting-edge method involves controlled electrical discharge in the presence of an insulating fluid, presenting a revolutionary approach to shaping Ni-Ti alloys. EDM technology, represented by systems like Hyflex EDM and Neoniti, stands at the forefront of endodontic innovation. These systems showcase not only enhanced durability and performance but also a commitment to precision and efficiency in root canal procedures. As endodontics continues to advance, EDM technology promises to play a pivotal role in shaping the future of root canal treatments.

In the late 1980s, ion implantation in plasma emerged as a groundbreaking technique. Applying a highly negative pulsating voltage to the plasma-submerged file in a vacuum chamber allowed ions (argon, boron, and nitrogen) to penetrate the file's surface without compromising its superelastic properties. Studies have demonstrated that ionic nitrogen implantation improves resistance to cyclic fatigue and enhances cutting efficiency, leading to improved wear resistance [63].

The immersion sol-gel method provides another avenue for enhancing file surfaces. By coating endodontic files with a protective layer of flexible TiO_2 through this method, a surface porous oxide film is formed. This film increases the stability of the surface layers, offering protection against corrosion [68].

The thermal nitriding method introduces titanium nitride (TiN) to the file's surface, consisting of a thin outer layer of TiN and a thicker inner layer of Ti_2Ni. This process significantly increases corrosion resistance, particularly in contact with sodium hypochlorite (NaOCl) which is used as an irrigant [68, 69].

Endodontic files subjected to these advanced surface treatments exhibit greater resistance to corrosive defects and improved overall performance. While some debate exists regarding the impact of file design versus finishing on cyclic fatigue resistance, it is evident that these surface treatments contribute significantly to the evolution of endodontic technology [69, 70].

4. Conclusion

Thanks to its biocompatibility, corrosion resistance, superelasticity, and fatigue resistance, Ni-Ti alloys are widely used in all areas of dentistry: prosthetics, orthodontics, and endodontics. The favorable characteristics include high compressive strength, similar to human bone, as well as an elasticity modulus nearly equivalent to bone tissue. These attributes are essential for the fabrication of dental implants.

Pseudoelasticity and shape memory effect are responsible for biomechanical compatibility. Thanks to its mechanical, antibacterial, and anticorrosive effect, this alloy is very successfully used for the production of orthodontic wires. Nickel-rich Ni-Ti alloys have attracted attention due to the possibility of stress-induced phase transformation, which is why most endodontic files rely on this feature. As technology moves forward, the integration of advanced surface treatments promises to further improve the efficiency and longevity of dental supplies, ultimately benefiting dental medicine as a whole.

Conflict of interest

The authors declare no conflict of interest.

Author details

Minja Miličić Lazić[1], Marijana Popović Bajić[2*], Igor Đorđević[1], Marija Živković[3], Vojkan Lazić[1], Vukoman Jokanović[4], Ilija Nasov[5] and Slavoljub Živković[2]

1 Department of Prosthodontics, School of Dental Medicine, University of Belgrade, Serbia

2 Department of Restorative Odontology and Endodontics, School of Dental Medicine, University of Belgrade, Serbia

3 Department of Orthodontics, School of Dental Medicine, University of Belgrade, Serbia

4 Institute for Nuclear Science, Vinča, Belgrade, Serbia

5 Plasma Center for Plasma Technology, Skopje, North Macedonia

*Address all correspondence to: dr.marijanapopovic@gmail.com

References

[1] Castleman LS, Motzkin SM, Alicandri FP, Bonawit VL, Johnson AA. Biocompatibility of nitinol alloy as an implant material. Journal of Biomedical Materials Research. 1976;**10**:695-731. DOI: 10.1002/jbm.820100505

[2] Genchi G, Carocci A, Lauria G, Sinicropi MS, Nickel CA. Human health and environmental toxicology. International Journal of Environmental Research and Public Health. 2020;**17**:679. DOI: 10.3390/ijerph17030679

[3] Shabalovskaya SA. On the nature of the biocompatibility and on medical applications of NiTi shape memory and superelastic alloys. Bio-medical Materials and Engineering. 1996;**6**:267-289

[4] Ryhänen J, Niemi E, Serlo W, Niemelä E, Sandvik P, Pernu H, et al. Biocompatibility of nickel-titanium shape memory metal and its corrosion behavior in human cell cultures. Journal of Biomedical Materials Research. 1997;**35**:451-457. DOI: 10.1002/(sici)1097-4636(19970615)35:4<451::aid-jbm5>3.0.co;2-g

[5] Nasakina EO, Sudarchikova MA, Sergienko KV, Konushkin SV, Sevost'yanov MA. Ion release and surface characterization of nanostructured nitinol during long-term testing. Nanomaterials (Basel). 2019;**9**:1569. DOI: 10.3390/nano9111569

[6] Duerig T, Pelton A, Stöckel D. An overview of nitinol medical applications. Materials Science and Engineering: A. 1999;**273-275**:149-160. DOI: 10.1016/S0921-5093(99)00294-4

[7] Soffa WA, Laughlin DE. Diffusional phase transformations in the solid state. In: Physical Metallurgy. Oxford, UK: Elsevier; 2014. pp. 851-1020. ISBN 978-0-444-53770-6

[8] Li G, Yu T, Zhang N, Chen M. The effect of Ni content on phase transformation behavior of NiTi alloys: An atomistic modeling study. Computational Materials Science. 2022;**215**:111804. DOI: 10.1016/j.commatsci.2022.111804

[9] Chekotu J, Groarke R, O'Toole K, Brabazon D. Advances in selective laser melting of nitinol shape memory alloy part production. Materials. 2019;**12**:809. DOI: 10.3390/ma12050809

[10] Sui S, Chew Y, Weng F, Tan C, Du Z, Bi G. Achieving grain refinement and ultrahigh yield strength in laser aided additive manufacturing of Ti–6Al–4V alloy by trace Ni addition. Virtual and Physical Prototyping. 2021;**16**:417-427. DOI: 10.1080/17452759.2021.1949091

[11] One Way and Two Way Shape Memory Effect: Thermo Mechanical Characterization of Ni-Ti Wires - PDF Free. Download Available from: https://docplayer.net/263607-One-way-and-two-way-shape-memory-effect-thermo-mechanical-characterization-of-ni-ti-wires.html [Accessed: July 2, 2023]

[12] Huang W, Toh W. Training two-way shape memory alloy by reheat treatment. Journal of Materials Science Letters. 2000;**19**:1549-1550. DOI: 10.1023/A:1006721022185

[13] Davis JR. Corrosion: Understanding the Basic (06691G). Philadelphia, USA: ASM International; 2020. ISBN 13: 9780871706416

[14] Cioffi M, Gilliland D, Ceccone G, Chiesa R, Cigada A. Electrochemical release testing of nickel-titanium orthodontic wires in artificial saliva using thin layer activation.

Acta Biomaterialia. 2005;**1**:717-724. DOI: 10.1016/j.actbio.2005.07.008

[15] Huang H-H, Chiu Y-H, Lee T-H, Wu S-C, Yang H-W, Su K-H, et al. Ion release from Ni-Ti orthodontic wires in artificial saliva with various acidities. Biomaterials. 2003;**24**:3585-3592. DOI: 10.1016/S0142-9612(03)00188-1

[16] Taqa A, Fathi W, Mohammed R. Evaluation of nickel ion release from orthodontic wires in different types of artificial saliva. RDENTJ. 2014;**14**:182-188. DOI: 10.33899/rden.2014.160888

[17] Lee T-H, Huang T-K, Lin S-Y, Chen L-K, Chou M-Y, Huang H-H. Corrosion resistance of different nickel-titanium archwires in acidic fluoride-containing artificial saliva. The Angle Orthodontist. 2010;**80**:547-553. DOI: 10.2319/042909-235.1

[18] O'Brien B, Carroll WM, Kelly MJ. Passivation of nitinol wire for vascular implants—A demonstration of the benefits. Biomaterials. 2002;**23**:1739-1748. DOI: 10.1016/S0142-9612(01)00299-X

[19] Eliaz N. Corrosion of metallic biomaterials: A review. Materials (Basel). 2019;**12**:E407. DOI: 10.3390/ma12030407

[20] Seo D-I, Lee J-B. Localized corrosion and repassivation behaviors of additively manufactured titanium alloys in simulated biomedical solutions. npj Materials Degradation. 2023;7:1-12. DOI: 10.1038/s41529-023-00363-4

[21] Bhandari J, Khan F, Abbassi R, Garaniya V, Ojeda R. Modelling of pitting corrosion in marine and offshore steel structures – A technical review. Journal of Loss Prevention in the Process Industries. 2015;**37**:39-62. DOI: 10.1016/j.jlp.2015.06.008

[22] Bae I, Kim B-H, Kim D-G, Sohn I-B, Yang S-W. Salt heat treatment and passivation to improve the corrosion resistance of nitinol (Ni-Ti). Materials (Basel). 2021;**14**:7789. DOI: 10.3390/ma14247789

[23] Modelling of Pitting Corrosion in Marine and Offshore Steel Structures – A Technical Review - ScienceDirect Available from: https://www.sciencedirect.com/science/article/abs/pii/S0950423015300024 [Accessed: January 17, 2024]

[24] Grgur B. Korozija I Zaštita. Belgrade, Serbia: Faculty of technology and metallurgy; 2020. ISBN: 978-86-7401-365-6

[25] Mohammadi Z, Soltani MK, Shalavi S, Asgary S. A review of the various surface treatments of Ni-Ti instruments. Iranian Endodontic Journal. 2014;**9**:235-240

[26] Sullivan SJL, Dreher ML, Zheng J, Chen L, Madamba D, Miyashiro K, et al. Effects of oxide layer composition and radial compression on nickel release in nitinol stents. Shape Memory and Superelasticity. 2015;**1**:319-327. DOI: 10.1007/s40830-015-0028-x

[27] Clarke B, Carroll W, Rochev Y, Hynes M, Bradley D, Plumley D. Influence of nitinol wire surface treatment on oxide thickness and composition and its subsequent effect on corrosion resistance and nickel ion release. Journal of Biomedical Materials Research. 2006;**79A**:61-70. DOI: 10.1002/jbm.a.30720

[28] Trepanier C, Venugopalan R, Pelton AR. Corrosion resistance and biocompatibility of passivated Ni-Ti. In: Yahia L, editor. Shape Memory Implants. Berlin, Heidelberg: Springer Berlin Heidelberg; 2000. pp. 35-45, ISBN 978-3-642-64118-3

[29] Shabalovskaya S, Anderegg J, Van Humbeeck J. Critical overview of nitinol surfaces and their modifications for

medical applications. Acta Biomaterialia. 2008;**4**:447-467. DOI: 10.1016/j.actbio.2008.01.013

[30] Arjmand F, Zhang L. Mechanical polishing, surface roughness, near-surface deformation, and electrochemical corrosion of alloy 690TT. Surface and Interface Analysis. 2015;**47**:1120-1126. DOI: 10.1002/sia.5858

[31] Milošev I, Kapun B. The corrosion resistance of nitinol alloy in simulated physiological solutions: Part 1: The effect of surface preparation. Materials Science and Engineering: C. 2012;**32**:1087-1096. DOI: 10.1016/j.msec.2011.11.007

[32] Rudolf R, Stambolić A, Kocijan A. Atomic layer deposition of $ATiO_2$ layer on Nitinol and its corrosion resistance in a simulated body fluid. Metals. 2021;**11**:659. DOI: 10.3390/met11040659

[33] Hanawa T, Ota M. Calcium phosphate naturally formed on titanium in electrolyte solution. Biomaterials. 1991;**12**:767-774. DOI: 10.1016/0142-9612(91)90028-9

[34] Metals as Biomaterials. Wiley. Available from: https://www.wiley.com/en-us/Metals+as+ Biomaterials-p-9780471969358 [Accessed: April 8, 2023]

[35] Shah KC, Chao D, Wu BM, Jensen OT. Shape-memory retained complete arch guided implant treatment using nitinol (Smileloc) abutments. Oral and Maxillofacial Surgery Clinics of North America. 2019;**31**:427-435. DOI: 10.1016/j.coms.2019.03.005

[36] Shah KC, Linsley CS, Wu BM. Evaluation of a shape memory implant abutment system: An up to 6-month pilot clinical study. The Journal of Prosthetic Dentistry. 2020;**123**:257-263. DOI: 10.1016/j.prosdent.2018.11.012

[37] Jokanović V, Petrović B, Zivkovic M. The main characteristics and application of the shape memory alloys in orthodontics and endodontics. Serbian Dental Journal. 2019;**66**:29-35. DOI: 10.2478/sdj-2019-0004

[38] Bahraminasab M, Sahari B. Ni-Ti Shape Memory Alloys, Promising Materials in Orthopedic Applications. London, UK: IntechOpen; 2013. pp. 261-278. ISBN 978-953-51-1084-2

[39] Choi J-W, Song C-H, Huh J-B. Implant-supported fixed dental prostheses with new retention type using zirconia ball and nickel-titanium spring. The Korean Academy of Oral & Maxillofacial Implantology. 2019;**23**:16-24. DOI: 10.32542/implantology.2019002

[40] Shah K, Seo Y, Wu B. Clinical application of a shape memory implant abutment system. The Journal of Prosthetic Dentistry. 2016;**117**:8-12. DOI: 10.1016/j.prosdent.2016.06.007

[41] Sachdeva R, Fukuyo S, Suzuki K, Oshida Y, Miyazaki S. Shape memory Ni-Ti alloys - applications in dentistry. Materials Science Forum. 1990;**56-58**:693-698. DOI: 10.4028/www.scientific.net/MSF.56-58.693

[42] Senthilnathan K, Shamimi A, Bonsignore C, Paranjape H, Duerig T. Effect of prestrain on the fatigue life of superelastic nitinol. Journal of Materials Engineering and Performance. 2019;**28**:5946-5958. DOI: 10.1007/s11665-019-04334-2

[43] Niinomi M, Liu Y, Nakai M, Liu H, Li H. Biomedical titanium alloys with Young's moduli close to that of cortical bone. Regen Biomater. 2016;**3**:173-185. DOI: 10.1093/rb/rbw016

[44] Melsen B, Cattaneo PM, Dalstra M, Kraft DC. The importance of force levels

in relation to tooth movement. Seminars in Orthodontics. 2007;**13**:220-233

[45] Fernandes DJ, Peres RV, Mendes AM, Elias CN. Understanding the shape-memory alloys used in orthodontics. ISRN Dentistry. 2011;**2011**:132408

[46] Otsuka K, Wayman CM, Nakai K, Sakamoto H, Shimizu K. Superelasticity effects and stress- induced martensitic transformations in CuAlNi alloys. Acta Metallurgica. 1976;**24**(3):207-226

[47] Harris DA, Jones AS, Darendeliler MA. Physical properties of root cementum: Part 8. Volumetric analysis of root resorption craters after application of controlled intrusive light and heavy orthodontic forces: A microcomputed tomography scan study. American Journal of Orthodontics and Dentofacial Orthopedics. 2006;**130**:639-647

[48] Sifakakis I, Bourauel C. Nickel–titanium products in daily orthodontic practice. In: Orthodontic Applications of Biomaterials. Cambridge, United States: Woodhead Publishing; 2017. pp. 107-127

[49] Bokas J, Woods M. A clinical comparison between nickel titanium springs and elastomeric chains. Australian Orthodontic Journal. 2006;**22**(1):39-46

[50] Mohammed H, Rizk MZ, Wafaie K, Almuzian M. Effectiveness of nickel-titanium springs vs elastomeric chains in orthodontic space closure: A systematic review and meta-analysis. Orthodontics & Craniofacial Research. 2018;**21**(1):12-19

[51] Shokeen B, Viloria E, Duong E, Rizvi M, Murillo G, Mullen J, et al. The impact of fixed orthodontic appliances and clear aligners on the oral microbiome and the association with clinical parameters: A longitudinal comparative study. American Journal of Orthodontics and Dentofacial Orthopedics. 2022;**161**(5):475-485

[52] Živkovic Sandic M, Popovic B, Carkic J, Nikolic N, Glisic B. Changes in subgingival microflora after placement and removal of fixed orthodontic appliances. Srpski Arhiv za Celokupno Lekarstvo. 2014;**142**(5-6):301-305

[53] Espinosa-Cristóbal LF, Holguín-MerázC, Zaragoza-ContrerasEA, Martínez-Martínez RE, Donohue-CornejoA, Loyola-RodríguezJP, et al. Antimicrobial and Substantivity properties of silver nanoparticles against oral microbiomes clinically isolated from young and young-adult patients. Journal of Nanomaterials. 2019;**2019**:3205971

[54] Jokanović V, Živković M, Zdravković N. A new approach to extraordinary efficient protection against COVID 19 based on nanotechnology. Serbia Dental Journal. 2020;**67**(2):100-109

[55] Gil FJ, Espinar-Escalona E, Clusellas N, Fernandez-Bozal J, Artes-Ribas M, Puigdollers A. New bactericide orthodonthic archwire: Ni-Ti with silver nanoparticles. Metals. 2020;**10**(6):702

[56] Shariat BS, Meng Q, Mahmud AS, Wu Z, Bakhtiari R, Zhang J, et al. Experiments on deformation behaviour of functionally graded Ni-Ti structures. Data in Brief. 2017;**13**:562-568

[57] Uysal I, Yilmaz B, Atilla AO, Evis Z. Nickel titanium alloys as orthodontic archwires: A narrative review. Engineering Science and Technology, an International Journal. 2022;**36**:101277

[58] Venkatesan K, Kailasam V, Padmanabhan S. Evaluation of titanium dioxide coating on surface roughness of nickel-titanium archwires and its

influence on Streptococcus mutans adhesion and enamel mineralization: A prospective clinical study. American Journal of Orthodontics and Dentofacial Orthopedics. 2020;**158**:199-208

[59] Sarakinos K, Alami J, Konstantinidis S. High power pulsed magnetron sputtering: A review on scientific and engineering state of the art. Surface and Coatings Technology. 2010;**204**:1661

[60] Tudose IV, Comanescu F, Pascariu P, Bucur S, Rusen L, Iacomi F, et al. Chapter 2 - chemical and physical methods for multifunctional nanostructured interface fabrication. In: Dinca V, Suchea MP, editors. Micro and Nano Technologies, Functional Nanostructured Interfaces for Environmental and Biomedical Applications. Elsevier; 2019. pp. 15-26. eBook ISBN: 9780128144022

[61] Ilić B, Petrović B, Marinković J, Miletić Vukajlović J, Stevanović M, Potočnik J, et al. Investigation of ion release and antibacterial properties of TiN-Cu-nanocoated nitinol archwires. Coatings. 2023;**13**:1587

[62] Zupanc J, Vahdat-Pajouh N, Schäfer E. New thermomechanically treated Ni-Ti alloys – A review. International Endodontic Journal. 2018;**51**:1088-1103. DOI: 10.1111/iej.12924

[63] Gavini G, Santos MD, Caldeira CL, Machado MEL, Freire LG, Iglecias EF. Nickel-titanium instruments in endodontics: A concise review of the state of the art. Brazilian Oral Research. 2018;**32**(Suppl. 1):e67

[64] Gambarini G, Cicconetti A, Di Nardo D, Miccoli G, Zanza A, Testarelli L. Influence of different heat treatments on torsional and cyclic fatigue resistance of nickel-titanium rotary files: A comparative study. Applied Science (Basel). 2020;**10**(16):5604-5611

[65] Gambarini G, Galli M, Di Nardo D, Seracchiani M, Donfrancesco O, Testarelli L. Differences in cyclic fatigue lifespan between two different heat treated Ni-Ti endodontic rotary instruments: Wave one gold vs edge one fire. Journal of Clinical and Experimental Dentistry. 2019;**11**(7):e609-e613

[66] Jamleh A, Alghaihab A, Alfadley A, Alfawaz H, Alqedairi A, Alfouzan K. Cyclic fatigue and torsional failure of edge taper platinum endodontic files at simulated body temperature. Journal of Endodontia. 2019;**45**(5):611-614

[67] Zinelis S, Eliades T, Eliades G. A metallurgical characterization of ten endodontic Ni-Ti instruments: Assessing the clinical relevance of shape memory and superelastic properties of Ni-Ti endodontic instruments. International Endodontic Journal. 2010;**43**(2):125-134

[68] Srivastava S, Alghadouni MA, Alotheem HS. Current strategies in metallurgical advances of rotary Ni-Ti instruments: A review. Journal of Dental Health, Oral Disorders & Therapy. 2018;**9**(1):72-77

[69] Han-Hsing Lin J, Karabucak B, Lee SM. Effect of sodium hypochlorite on conventional and heat-treated nickel-titanium endodontic rotary instruments - an *in vitro* study. Journal of Dental Science. 2021;**16**(2):738-743

[70] Huang X, Shen Y, Wei X, Haapasalo M. Fatigue resistance of nickel-titanium instruments exposed to high-concentration hypochlorite. Journal of Endodontia. 2017;**43**(11):1847-1851

Chapter 6

Titanium-Based Alloys: Classification and Diverse Applications

Nada H.A. Besisa and Takeaki Yajima

Abstract

Titanium-based alloys have emerged as pivotal materials across numerous industries due to their exceptional properties, including high strength-to-weight ratios, corrosion resistance, and biocompatibility. This chapter provides a comprehensive overview of the classification and diverse applications of titanium-based alloys, spanning aerospace, medical implants, automotive engineering, and beyond. Through case studies and technological advancements, the chapter elucidates the remarkable history of titanium alloys and their contributions to innovation, sustainability, and enhanced performance in various sectors. Special attention is given to Ti-6Al-4V, a widely utilized alloy renowned for its unique properties. Overall, this chapter offers insights into the widespread influence and promising future prospects of titanium-based alloys in shaping modern technological landscapes.

Keywords: biomaterials, aerospace, industries, construction, automobile

1. Introduction

Titanium, a versatile and robust metal, has been a cornerstone in various industries for over two centuries since its discovery by British mineralogist William Gregor in 1791. The development of Ti-6Al-4V, one of its most popular alloys, occurred successfully in the 1940s. Renowned for its exceptional properties such as high corrosion resistance, remarkable strength-to-weight ratio, and biocompatibility, titanium, and its alloys have found widespread applications across sectors ranging from aerospace to medical, chemical processing, offshore and marine engineering, power generation, medicine, transportation, architecture, and consumer goods [1–4]. With a density approximately 60% lower than that of steel and superalloys, titanium exhibits remarkable lightweight properties. In addition to their remarkable resistance to corrosion, titanium and its alloys exhibit exceptional properties such as high fracture toughness, high-temperature strength, and an impressive strength-to-weight ratio [1]. Titanium alloys, despite being 45% lighter than standard low-carbon steels, surpass them in strength. They are only 60% heavier but twice as strong as soft aluminum alloys. Moreover, through alloying and deformation processing of Ti-alloys, substantial enhancements in strength can be achieved [1]. In this chapter, we will delve into the classification of titanium-based alloys, which is determined by both chemical composition and thermomechanical processing [2].

In addition, we will explore their diverse applications in detail, highlighting key examples and considerations.

2. Classification of titanium-based alloys

Titanium is available as commercially pure and as alloys. Pure titanium, in its elemental form, exhibits characteristics such as low thermal conductivity, relatively low density and elastic modulus, moderate strength, excellent corrosion resistance in diverse environments, and high reactivity with various elements. **Table 1** provides a comparison of selected properties [2] of titanium with those of competing metals, where titanium shows outstanding properties among others. The microstructure and properties of these alloys are influenced by factors such as chemical composition and thermomechanical processing. At low temperatures, pure titanium adopts a hexagonal close-packed structure (hcp), known as α-titanium. However, at elevated temperatures, it transitions to a body-centered cubic (bcc) structure, referred to as β-titanium. **Figure 1** shows the atomic unit cells of these structures. The β-transus temperature for pure titanium is approximately 882°C [5], which can vary with the presence of incorporated impurities [6]. The coexistence of these two crystal structures forms the basis for the diverse range of properties observed in titanium alloys. The alloying elements utilized in titanium alloys are categorized as neutral,

	Ti	Al	Ni	Fe
Density, g/cm^3	4.5	2.7	8.9	7.9
Melting point, °C	1670	660	1455	1538
Thermal conductivity, W/mK	15–22	221–247	72–92	68–80
Elastic modulus, GPa	115	72	200	215
Reactivity with oxygen	High+	High	Low	Low
Corrosion resistance	High+	High	Medium	Low
Cost	High+	High	High	Low

Table 1.
Physical properties of titanium and other selected contestant materials.

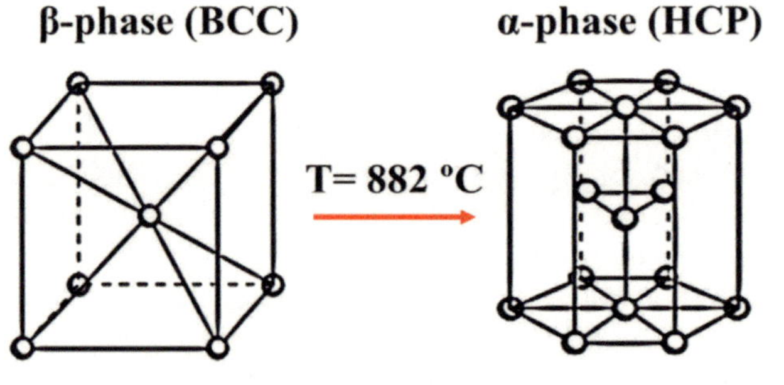

Figure 1.
The crystalline structure and phase transformation of elemental titanium.

	α-Stabilizer				β-Stabilizer							Neutral	
					β-Eutectoid		β-Isomorphous						
	Al	O	N	C	Mo	V	Fe	Cr	Mn	H	Ni	Sn	Zr
Substitutional	o				o	o						o	o
Interstitial		o	o	o			o	o	o	o	o		

Table 2.
Alloying elements used in titanium-based alloys.

α-stabilizers, or β-stabilizers [5–7] based on their impact on stabilizing α or β phases (as shown in **Table 2**). This table also delineates the position of the alloying element within the lattice crystal, which can be either interstitial or substitutional. α-stabilizers elevate the β-transus temperature, whereas β-stabilizers decrease it. Neutral elements exert minimal influence on the β-transus temperature. β-Stabilizers are further classified into β-isomorphous and β-eutectoid elements. The former promotes β phase stability across all alloy compositions, while the latter induces eutectoid transformations of the β phase [2]. Notably, aluminum serves as the primary α-stabilizer, while molybdenum ranks among the principal β-stabilizers. The α-stabilizers and their relative efficacy in stabilizing the α phase are quantified in terms of aluminum equivalence, with molybdenum holding similar significance for the β phase [8].

According to their metallurgical structure, titanium-based alloys can be categorized into three main groups: alpha (α), alpha-beta (α-β), and beta (β). In addition, titanium-based alloys are subdivided into near alpha and metastable beta alloys (see **Table 3**) [1, 2]. Alpha alloys are primarily composed of α-phase structures, which include both pure titanium and alloys infused with α-stabilizers like aluminum and tin. These alloys are frequently utilized in aerospace applications due to their distinct

Category	Examples
Alpha alloys	Commercially pure titanium—ASTM Grades 1, 2, 3, and 4 Ti/Pd alloys—ASTM Grades 7 and 11 Ti-2Cu
Near alpha alloys	Ti-8Al-1Mo-1V Ti-6Al-5Zr-0.5Mo-0.2Si-IMI 685 Ti-6Al-4Zr-3Sn-2Mo-0.08Si Ti-5.5Al-3Zr-3.5Sn-0.3Mo-1Nb-0.3Si-IMI 829
Alpha-Beta alloys	Ti-6Al-4V Ti-6Al-6V-2Sn Ti-4Al-4Mo-4Sn-0.5Si Ti-6Al-2Sn-4Zr-6Mo
Beta alloys	Ti-13V-11Cr-3Al T-13V-11Cr-3Al Ti-11.5Mo-6Zr-4.5Sn
Metastable beta alloys	Ti-3Al-8V-6Cr-4Mo-4Zr-Beta C Ti-6V-6Mo-5.7Fe-2.7Al-TIMETAL 125 Ti-15V-3Cr-3Sn-3Al

Table 3.
Classification of titanium-based alloys with examples.

characteristics, which balance strength and formability [9]. Notable examples of alpha alloys include ASTM Grades 1–4 and Ti/Pd alloys (ASTM Grades 7 and 11). Near alpha alloys exhibit a predominant α-phase with limited β-stabilizers, striking a balance between strength and formability, making them adaptable for various applications [10].

Alpha-beta alloys maintain an equilibrium between α and β phases, offering a blend of strength, ductility, and heat resistance that finds extensive usage across diverse industries [9]. Examples of alpha-beta alloys include the popular Ti-6Al-4V and Ti-6Al-6V-2Sn, known for their exceptional properties and widespread availability. Beta alloys are primarily composed of the β phase and are characterized by elements such as vanadium and molybdenum [11]. These alloys prioritize ductility over mechanical strength compared to alpha-beta alloys and find applications in specialized industries. Notable examples include Ti-13V-11Cr-3Al and Ti-11.5Mo-6Zr-4.5Sn. Beta metastable alloys, or near beta alloys, are chiefly composed of β phases with restricted α-stabilizers, prioritizing ductility over mechanical strength compared to α-beta alloys [12]. Moreover, Nitinol, although technically classified as a nickel-titanium intermetallic, is included among titanium alloys due to its widespread application in biomedical fields despite its high nickel content [13]. However, as it is an intermetallic rather than a traditional alloy, it will not be extensively discussed in this chapter. Additionally, **Table 3** presents examples of alloys with varying chemical compositions, highlighting both underutilized and highly sought-after alloys like Ti-6Al-4V, renowned for its unique properties and significant demand across numerous industries. Understanding the classification and characteristics of these alloys is essential for selecting the appropriate material for specific industrial needs.

3. Applications of titanium-based alloys

Among the categories of titanium-based alloys, the α + β alloys hold the largest share at 70% in the US market. Globally, Ti-6Al-4V constitutes over 50% of titanium alloy consumption, while commercially pure titanium accounts for approximately 20–30% [5]. Although there are more than 100 known titanium alloys, only 20–30 have attained commercial status. Recently, there has been growing interest in titanium aluminides, particularly γ(TiAl)-based alloys, for aerospace and automotive applications.

Titanium and its alloys find extensive use across various industries, with selection criteria often based on corrosion resistance or strength requirements. Biocompatibility is also a critical consideration for biomedical implant applications. Commercially pure Ti (ASTM Grades 1–4) is commonly employed for corrosion-resistant applications due to its good corrosion resistance but relatively low strength. Grades 7, 8, and 11 are utilized for specific corrosion resistance needs. In the medical field, Grade 2 is preferred for low-strength applications, while Grade 5 (Ti-6Al-4V) is chosen for higher strength requirements [14]. For applications demanding high strength, titanium alloys like Ti-6Al-4V, Ti-8Al-1Mo-1V, Ti-6Al-2Sn-4Zr-2Mo, Ti-6Al-6V-2Sn, and Ti-10V-2Fe-3Al are utilized. Among these, Ti-6Al-4V stands out due to its unique combination of properties, workability, widespread production experience, and commercial availability. Consequently, it has become the benchmark against which other titanium alloys are compared when selecting for specific applications [14]. Let us explore some of the key applications of titanium-based alloys in detail:

3.1 Applications of titanium-based alloys in aerospace Industry

Titanium-based alloys have garnered significant attention and application in the major field of aerospace industry due to their exceptional properties, where their lightweight yet strong characteristics are highly valued, making them a preferred choice for various critical components. From structural components to jet engine parts and spacecraft components, these alloys contribute to enhanced performance, efficiency, and reliability in aerospace applications. This section explores the diverse applications of titanium alloys within aerospace, focusing on their structural, engine, fastener, and spacecraft applications [15, 16].

3.1.1 Structural components

Titanium alloys, notably Ti-6Al-4V, are extensively used in aerospace for structural components. Their high strength-to-weight ratio and corrosion resistance make them ideal for airframe structures, landing gear components, and other critical load-bearing parts in aircraft [17].

3.1.2 Jet engine components

In the realm of jet engines, titanium-based alloys find indispensable use. Their ability to withstand high temperatures and aggressive environments makes them suitable for turbine blades, discs, and casings, contributing to increased efficiency and reliability of jet engines [18].

3.1.3 Aerospace fasteners

Titanium alloys are extensively employed in aerospace fasteners due to their high strength, low weight, and excellent corrosion resistance. These alloys play a crucial role in securing critical components while minimizing overall weight, enhancing fuel efficiency, and ensuring structural integrity [19].

3.1.4 Spacecraft and satellite components

The lightweight nature and durability of titanium-based alloys make them indispensable in spacecraft and satellite construction. From structural components to thermal shields and satellite frames, these alloys ensure reliability and endurance in the harsh conditions of space [20].

The unique combination of properties possessed by titanium-based alloys has positioned them as vital materials in the aerospace industry. Their contributions to structural integrity, engine efficiency, fastening systems, and space exploration continue to drive advancements in aerospace technology [21].

Table 4 outlines various applications of titanium materials in aerospace, highlighting their advantages over aluminum alloys in terms of higher strength leading to weight savings. Titanium alloys are often substituted for aluminum alloys in areas where operational temperatures exceed the limits of aluminum. These areas include nacelles, auxiliary power units, and wing anti-icing systems. For instance, landing gear beams on aircraft like the Boeing 747 and 757 demonstrate the challenge of volume constraints, which can be addressed by utilizing

Material	Application	
Commercially pure titanium	Airframe structure	Floors
Ti-6Al-4V		Windows frames
Ti-10V-2Fe-3Al; Ti-6-6-2		Landing gear
Ti-3Al-2.5V		Hydraulic tubing
Ti-15V-3Cr-3Sn-3Al		Springs
Ti-6Al-4V; Ti-6-2-4-2S	Gas turbine engines	Compressor disc
		Compressor blades
		Fan discs and blades
Ti-35V-15Cr		Compressor stators
TIMETAL21S		Nozzle assembly

Table 4.
Applications of selected titanium-based alloys in the aerospace industry.

titanium alloys despite their higher cost compared to aluminum. Titanium's corrosion resistance obviates the need for painting in most cases, except when galvanic corrosion risk arises from contact with aluminum or low alloy steel components. For structures exposed to highly corrosive environments, such as the floor support under kitchens and lavatories, titanium ensures better structural durability. Some examples of commonly used titanium-based alloys in airframe structure such as floors, windows frames, landing gears and springs are: commercially pure titanium, Ti-6Al-4V, Ti-10V-2Fe-3Al, Ti-6-6-2, and Ti-15V-3Cr-3Sn-3Al. On the other hand, Ti-6Al-4V, Ti-6-2-4-2S, Ti-35V-15Cr, and TIMETAL21S are commonly used in parts of gas turbine engines such as compressor discs, compressor blades, fan discs and blades, compressor stators, and nozzle assembly. Moreover, **Table 5** provides further details on the application of titanium materials in aerospace, categorized by alloy type, reinforcing the versatility and importance of titanium in this industry.

3.2 Applications of titanium-based alloys in medical devices and implants

In comparison to conventional stainless steel and Co-Cr alloys, the biocompatibility and corrosion resistance of titanium alloys make them ideal for medical implants and devices. Orthopedic implants like joint replacements, bone plates, and dental implants often use titanium alloys due to their ability to integrate well with human tissues and withstand the body's harsh physiological environment [22, 23]. Utilizing a broad spectrum of available alloys, the biomedical industry predominantly relies on commercially pure titanium Grade 2 and Ti-6Al-4V Grade 5 for over 95% of titanium biomedical devices. Additionally, ELI (extra-low interstitials) alloys are prevalent in biomedical applications, boasting a chemical composition akin to the aforementioned alloys but with significantly reduced interstitial element levels. The diminished presence of interstitial elements, namely oxygen, nitrogen, hydrogen, and boron, within the alloy confers advantageous enhancements in material ductility and fracture toughness [24]. Various titanium alloys are employed in biomedical devices based

Alloy type		Application
α Alloy	Commercially pure titanium	The annealed conditions for floor support structure in the areas of galley and lavatories. Brackets and clips. Pipes/tubes in the lavatory system. Ducting for the anti-icing. Environmental control systems at temperatures up to 230°C.
	Ti-5-2.5	The hydrogen side of the high-pressure fuel turbo-pump of the space shuttle. The annealed condition for cryogenic applications.
	Ti-6-2-4-2S	Parts of gas turbine engine at temperatures up to 540°C.
	Timetal-II00 (Ti-6Al-2.8Sr-4Zr-0.4Mo-0.4Si)	Allison gas turbine engines, at temperatures up to 600°C.
	Ti-3Al-2.5V	High pressure hydraulic lines. Fabrication of honeycomb core.
	Ti-8-1-1	Fan blades for military engines. Tear straps on commercial airframes.
α + β Alloy	Ti-6Al-4V	Static and rotating components of gas turbine engines covering all sections of aircraft. The floor support structure in galleys and lavatory areas.
	Ti-6-2-4-6	Moderate temperatures applications up to 315°C. Military engines, for example F-119 and F-100.
	Ti-5Al-2Sn-2Zr-4Mo-4Cr	Fan and compressor discs below 400°C.
β Alloy	Ti-6Al-2Sn-2Zr-2Mo-2Cr + Si (Ti-6-22-22)	The Lockheed/Boeing F-22 program is as moderate strength-damage tolerant alloy.
	Ti-13V-11Cr-3Al (Ti-13-11-3)	Wing and body skins, bulkheads, rivets, frames, ribs, and landing gears of SR-71 airplane.
	Ti-10-2-3	The entire main landing gear of the 777.
	Ti-15-3	Springs from flat products, such as clock-type springs. Strip is its primary product form.
	Alloy C	Exhaust structure and cast compressor components of F-119 engine that powers the Lockheed/Boeing F-22.

Table 5.
Applications of selected titanium-based alloys in the aerospace industry based on alloy type.

on their specific function, size, shape, and anatomical location. This section will delve into the diverse biomedical applications, with a comprehensive list provided in **Table 6** [25–31].

3.2.1 Cardiovascular devices

Titanium alloys are also employed in cardiovascular devices such as stents, pacemaker cases, and heart valve components. Their biocompatibility and resistance to corrosion in bodily fluids make them suitable for these critical applications within the cardiovascular system [32]. Titanium and its alloys are integral to the advancement of cardiovascular devices, playing a pivotal role in enhancing patient outcomes for heart

Material	Application
Commercially pure titanium, Ti-6Al-4V, Ti-6Al-7Nb, Ti-15Mo, Nitinol	Cardiovascular devices (heart connectors, valves, catheters, implantable defibrillators, ventricular assist devices)
Commercially pure titanium, Ti-6Al-4V, Ti-6Al-7Nb, Ti-15Mo, Ti-13Nb-13Zr, Nitinol	Orthopedic implants (hip and knee joints, meshes, bone substitute, fixation devices)
Commercially pure titanium (grades 1, 2, 3, and 4), Ti-6Al-4V, –titanium, Ti, Nitinol	Dental implants (braces, bridges, fixation devices, abutments)
Commercially pure titanium, Ti-6Al-4V, Ti-6Al-7Nb	trauma devices (screws, plates, nails, nodes)
	Soft tissue implants (breast reconstruction meshes, hernia meshes, fixation devices)

Table 6.
Medical applications of some titanium-based alloys.

and vascular conditions. These alloys exhibit properties that make them highly suitable for devices aimed at restoring normal blood flow, improving cardiac function, and providing structural support [9].

In cardiovascular applications, titanium alloys are utilized in various devices, including coronary and peripheral vascular stents, which are employed to open narrowed or blocked arteries, thus restoring blood flow and preventing complications like heart attacks. Additionally, titanium alloys are used in artificial mechanical heart valves, replacing damaged tissue to ensure proper blood flow through the heart chambers.

The biocompatibility and corrosion resistance of titanium makes it an ideal material for enclosures housing pacemakers and implantable cardioverter-defibrillators (ICDs), safeguarding sensitive electronics from the harsh biological environment and external electromagnetic interferences. Unlike materials used in other medical devices, the choice of titanium for pacemakers and defibrillators is primarily driven by its chemical resistance and insulation capabilities rather than mechanical strength. Titanium can also be employed for electrode tips in these devices [33].

The mechanical properties of titanium alloys are crucial for cardiovascular applications, particularly in stents and mechanical heart valves. Stents, which maintain the openness of narrowed or blocked blood vessels, are typically made from metals such as titanium, stainless steel, platinum-iridium alloys, tantalum, and cobalt-chromium alloys. To prevent restenosis, stents are often coated with hard and anti-adhesive layers like titanium oxide and titanium nitride [9, 34].

Similarly, heart valves, subjected to mechanical loads, are coated with anti-adherent layers to prevent cellular adhesion and obstruction. Titanium alloys like Ti-6Al-4V, commercially pure titanium (Grade 2), and Nitinol find application in cardiovascular devices due to their mechanical strength, corrosion resistance, biocompatibility, and unique properties such as shape memory and super elasticity [9, 35, 36].

3.2.2 Orthopedic implants

Titanium and its alloys are instrumental in addressing the intricate requirements of musculoskeletal devices, ranging from joint replacements to fixation components [9, 37]. These materials have revolutionized the development of implants, seamlessly integrating with bone and significantly enhancing mobility, function, and overall quality of life for countless patients [28, 38]. Notably, titanium's remarkable strength enables the fabrication of durable implants capable of withstanding physiological loads, ensuring stability and longevity in orthopedic settings.

While cobalt-chromium alloys have historically achieved similar outcomes in orthopedics, titanium offers two critical advantages: easier osseointegration and a higher strength-to-weight ratio, resulting in lighter implants [9]. Although titanium is utilized in various joint implants, it is never used as an articulating component due to limitations in wear and tribo-corrosion resistance. Instead, titanium excels as a material for load-bearing components, such as femoral stems and acetabular cups in hip replacements, providing essential stability and support to transfer mechanical loads to surrounding bone [9].

Beyond joint prostheses, titanium finds application in nonarticulating bone implants like ribcages, skull implants, spinal cages, and bone scaffolds, where minimal risk of tribological damage exists [39, 40]. These fixed devices benefit from titanium's longevity and reliability, as they lack moving parts or surfaces in contact, reducing wear, friction-induced debris, and corrosion.

Several titanium alloys are approved for orthopedic use, including Ti-6Al-4V, Ti-6Al-7Nb, Ti-15Mo, Ti-13Nb-13Zr, and commercially pure titanium [9]. Each alloy offers unique properties suited for specific applications, such as load-bearing implants, spinal implants, bone screws, and fixation devices. Despite concerns like ion release and stress shielding, titanium alloys have demonstrated remarkable clinical success, with some implants lasting over three decades in challenging environments.

3.2.3 Dental prosthetics

The biocompatibility and corrosion resistance of titanium-based alloys makes them ideal for dental implants and prosthetics. These alloys are used in dental implants, crowns, bridges, and orthodontic appliances, offering durability and compatibility with oral tissues [41]. The utilization of titanium has brought about a transformative impact on both dental implants and orthodontic braces, leading to significant enhancements in patient outcomes and comfort [42]. Dr. Per-Ingvar Brånemark's discovery of osseointegration, the direct bonding between bone and titanium surfaces, laid the foundation for titanium's widespread use in dentistry and orthodontics [9].

Dental implants, particularly titanium posts, have become the preferred method for replacing missing teeth. Surgically placed into the jawbone, these posts gradually integrate with surrounding bone tissue, providing a stable foundation for prosthetic teeth. Dental implants not only restore chewing and speech functions but also prevent bone loss, preserving facial structure and overall oral health [43].

In orthodontics, titanium alloys are instrumental in creating braces, wires, and other devices due to their exceptional strength-to-weight ratio and corrosion resistance. These properties enable the application of controlled forces to move teeth into proper alignment, leading to effective and predictable outcomes [25].

The oral environment poses unique challenges for dental implants and orthodontic devices, including pH fluctuations and bacterial colonization. Titanium's corrosion resistance and biocompatibility are critical in combating these challenges, ensuring the longevity and stability of dental devices.

Long-term clinical studies affirm the success of titanium-based dental devices, with impressive survival rates exceeding 95% over 10 years [26, 27]. The most common titanium alloys in dentistry include β-titanium alloys, Ti-6Al-4V, commercially pure titanium, and Nitinol, each catering to specific dental applications ranging from wires and brackets to bone plates and orthodontic archwires [9].

3.2.4 Surgical and trauma instruments

The strength, durability, and corrosion resistance of titanium alloys make them ideal for surgical instruments. These instruments include forceps, retractors, and scalpels, benefiting from the lightweight yet robust nature of titanium-based materials [44].

The most used titanium materials in surgical devices are mainly Ti-6Al-4V and commercially pure titanium, where the latter was considered to be the best biocompatible metallic material due to its surface properties enabling the spontaneous build-up of stable and inert oxide layer [45, 46]. Trauma devices, including bone plates, screws, and intramedullary nails, necessitate materials capable of withstanding significant mechanical stresses within the human body, often surpassing those experienced by orthopedic implants. Titanium, alongside cobalt-chromium and stainless steel, is commonly used in trauma devices despite being less mechanically robust, particularly under cyclic fatigue conditions [9]. Research suggests that titanium implants may have higher failure rates due to fracture compared to stainless steel alternatives, but they offer superior biological properties and lower infection risks [47]. However, the design of titanium implants must carefully consider anatomical location, as titanium's softness can lead to the production of cytotoxic particulates through abrasive wear [9, 48, 49].

In screw designs, while thread fractures are rare, shaft fractures are relatively common, indicating weaker forces at the thread-bone interface compared to bending stresses on the shaft [50]. Additionally, the stress distribution differs significantly between locking and conventional plates. Locking plates, which anchor screws directly to the plate, offer enhanced stability but may increase mechanical stress on both bone and device, potentially leading to tissue damage [51, 52]. Despite these considerations, titanium alloys such as Ti-6Al-4V, Ti-6Al-7Nb, and commercially pure titanium remain the primary choices for trauma devices.

3.2.5 Challenges and considerations

However, Ti-6Al-4V offers excellent properties for many applications. Its relatively poor wear resistance makes it unsuitable for bearing surface applications without surface treatments [45].

Recent research has indicated that the elastic behavior of $\alpha + \beta$ titanium alloys like Ti-6Al-4V may not be ideal for orthopedic applications due to their mismatch with the elastic modulus of cortical bone [2]. This can lead to inadequate load transfer to the adjacent bone and subsequent degradation [53]. Furthermore, the presence of toxic elements like vanadium (V) and aluminum (Al) in Ti-6Al-4V alloy has raised concerns about biocompatibility. As a result, vanadium-free alloys such as Ti-6Al-7Nb and Ti-5Al-2.5Fe have been developed, offering improved biocompatibility, higher fatigue strength, and lower elastic modulus [45].

Efforts to develop alloys completely free of toxic elements have led to the creation of alloys like Ti-13Nb-13Zr, Ti-12Mo-6Zr-2Fe, and Ti-29Nb-13Ta-4.6Zr [54]. Authors suggest that suitable heat treatment and the addition of biocompatible alloying elements such as niobium (Nb), tantalum (Ta), and zirconium (Zr) to titanium are essential for achieving titanium alloys with optimal mechanical properties, including low elastic modulus and excellent biocompatibility [53, 55].

The versatility and biocompatibility of titanium-based alloys continue to drive innovation in medical device manufacturing. Ongoing research focuses on enhancing surface modifications and biodegradability and integrating advanced technologies to further improve the performance of these materials in medical applications [56]. To address certain challenges associated with conventional titanium alloys, Haase et al. [57] have developed two novel alloys, Ti-0.44O-0.5Fe-0.08C-0.4Si-0.1Au and Ti-0.44O-0.5Fe-0.08C-2.0Mo, characterized by medium to high strength. Notably, these alloys exclusively incorporate elements either found naturally in the human body or known to be biocompatible. Their study demonstrated exceptional mechanical properties, suggesting their potential as viable alternatives to Ti-6Al-4V for medical applications.

Titanium-based alloys have revolutionized the medical field, offering a unique combination of biocompatibility, strength, and corrosion resistance. Their applications in orthopedic implants, dental prosthetics, cardiovascular devices, and surgical instruments continue to improve patient outcomes and drive advancements in medical technology [23].

3.3 Applications of titanium-based alloys in automotive and marine applications

In the automotive industry, titanium alloys are used in exhaust systems, engine components, and other parts where high-temperature resistance, strength, and corrosion resistance are crucial. Similarly, in marine applications, titanium alloys' resistance to corrosion in saltwater environments makes them suitable for components such as propeller shafts and hulls [58, 59].

3.3.1 Lightweight components

Titanium-based alloys contribute to reducing vehicle weight in both automotive and marine applications. Components such as valves, connecting rods, and exhaust systems benefit from the high strength-to-weight ratio of titanium alloys, enhancing performance and fuel efficiency [60].

3.3.2 Engine systems

In high-performance engines, titanium alloys are used in valve systems due to their excellent heat resistance, corrosion resistance, and strength at elevated temperatures. Titanium valves offer improved engine performance, reliability, and durability, especially in racing and high-end automotive and marine engines [61]. Ti-6Al-4V, γ(TiAl), Grade 2, and Ti-6Al-2S-4Zr-2Mo-0.1Si are commonly used alloys in parts of the engine [2, 61].

3.3.3 Corrosion-resistant parts

Titanium-based alloys are employed in marine environments for their exceptional corrosion resistance. Components exposed to saltwater, such as propeller shafts, hulls, and fasteners, benefit from the anti-corrosive properties of titanium alloys, extending their lifespan in marine applications [62].

3.3.4 Suspension systems

In automotive applications, titanium alloys such as Ti-6Al-4V, and Ti-6.8Mo-4.5Fe-1.5Al are utilized in suspension systems due to their high strength and corrosion resistance. Suspension components made from these alloys offer durability and enhanced performance under various road conditions [63].

3.3.5 Challenges and considerations

Titanium-based alloys play a pivotal role in automotive and marine applications, offering a balance of strength, corrosion resistance, and lightweight properties. Their use in lightweight components, engine systems, corrosion-resistant parts, and suspension systems contributes to improved performance, efficiency, and durability in vehicles and marine vessels [60]. However, the high cost of titanium alloys has historically limited their use in automobiles to racing and specialized vehicles, despite the automotive industry's interest in their lightweight properties, fuel efficiency, and performance benefits. On the other hand, in recent years, there has been a growing adoption of titanium and its alloys for various automobile components (see **Table** 7) [2, 64]. A significant number of titanium intake valves, primarily made of Ti-6Al-4V alloy, have been installed in many cars and motorcycles [64]. Surface treatment has been a key challenge, with efforts focused on improving wear resistance through methods such as TiN coating, Mo thermal spray coating, and Cr plating [2]. These treatments, however, are costly and not always effective for prolonged wear resistance. As an alternative, an oxidizing treatment has been developed to enhance hardness by forming a thick, hardened layer through the diffusion of concentrated oxygen into the titanium surface layer.

For exhaust valves, which endure high temperatures, the heat-resistant alloy Ti-6Al-2Sn-4Zr-2Mo-0.1Si (6242S) is commonly used. However, for mass-produced

Material	Application	
Ti-6Al-4V	Engine	Intake valve Connecting rods Outlet valves
γ(TiAl)		Turbocharger rotors Outlet valve
Grade 2		Exhaust system Outlet valve
Ti-6Al-2S-4Zr-2Mo-0.1Si		Outlet valve
Ti-6Al-4V	Frame structure	Body Armor Suspension springs
Commercially pure titanium (Grade 4)		Body
Ti-6.8Mo-4.5Fe-1.5Al		Suspension springs

Table 7.
Automotive applications of some titanium-based alloys.

motorcycles subjected to even higher temperatures for extended periods, research has explored the application of TIMETAL@1100 (Ti-6Al-2.7Sn-4Zr-0.4Mo-0.45Si), one of the most heat-resistant titanium alloys available. It was found that this alloy's service temperature aligns with the requirements for motorcycle exhaust valves [2].

Table 8 outlines the adoption of titanium alloys by automakers in their industries, with weight reduction being a primary focus and benefit. Automobile manufacturers like Mitsubishi, Honda Motors, Toyota, and Nissan Motors adapted titanium alloys such as Ti-22V-4Al, Ti-6Al-4V, Ti-6Al-2S-4Zr-2Mo-Si, Ti-3Al-2.5V+REM, Ti-Al-Zr-Sn-Nb-Mo-Si/TiB, Ti-6Al-4V/TiB, and Ti-4.5Fe-6.8Mo-1.5Al that were introduced beginning in 1989 [64]. Ongoing research is also investigating new alloys for automobile parts, including Super-TIX, Super-TIX51AF (Ti-5%Al-1%Fe), Super-TIX800 (Ti-1%Fe-0.35%O-0.01%N), and TIMETAL@LCB (Ti-4.5Fe-6.8Mo-1.5Al), among others [64].

3.4 Applications of titanium-based alloys in sports equipment and consumer goods

The high strength-to-weight ratio of titanium alloys makes them attractive for sporting equipment like bicycle frames, golf clubs, and tennis rackets. Additionally, due to their esthetic appeal, corrosion resistance, and durability, titanium alloys are utilized in luxury goods such as watches, jewelry, and eyewear [65, 66].

3.4.1 Sports equipment

Titanium-based alloys offer a unique combination of strength and lightness, improving performance and endurance in sports gear. Titanium frames in bicycles, for example, provide high strength while maintaining a lightweight structure, enhancing maneuverability and speed [67]. The utilization of titanium in sports equipment has advanced significantly, ranging from early tennis and badminton rackets to modern golf heads, handles, racing bicycles, and even racing cars, marking a notable progression in the comprehension of titanium's capabilities [67].

In golf, titanium's lightweight and high strength properties have allowed for the creation of larger club heads without increasing overall weight, resulting in improved performance and distance for golfers. Innovative titanium alloys, such as those

Material	Producer	Application
Ti-22V-4Al	Mitsubishi	AMG engine retainers of the Gallant 1
Ti-6Al-4V Ti-6Al-2S-4Zr-2Mo-Si	Nissan Motors	Engine inlet and exhaust valves for the CIMA
Ti-3Al-2.5V + REM	Honda Motors	Connecting rods of sport car NSX
Ti-Al-Zr-Sn-Nb-Mo-Si/TiB Ti-6Al-4V/TiB	Toyota	Intake and exhaust engine valves in the Altezza
Ti-4.5Fe-6.8Mo-1.5Al	Volkswagen	Suspension spring of Lupo FS
Titanium alloys	Kawasaki General Motors	Muffler of the large sports-type motorcycle ZX-9 Dual mufflers of the Corvette Z06

Table 8.
Automotive applications and producers of some titanium-based alloys.

developed by Nippon Kokan KK, offer enhanced durability for golf head surfaces, contributing to their popularity in the market.

In tennis and badminton, the incorporation of titanium components, such as pure titanium nets and super-elastic titanium-nickel alloy handles, has bolstered racket performance, leading to increased hitting power and user satisfaction. Additionally, ongoing research into titanium fiber materials aims to leverage titanium's rebound force to further enhance racket effectiveness [68].

Racing bicycles benefit from titanium's lightness and strength, with titanium components reducing weight and wind resistance. Titanium bicycle frames, composed of industrial pure titanium tubes and sport-grade titanium alloys, have gained popularity among cyclists worldwide, especially in high-end bicycle sports [68].

In the realm of racing cars, titanium's exceptional physical and mechanical properties have led to its widespread use in various components, including bolts, connecting rods, exhaust pipes, and brakes. The application of titanium contributes to reduced vehicle weight, lower fuel consumption, and improved environmental impact, showcasing titanium's versatility and effectiveness in automotive engineering [68].

Furthermore, titanium finds applications in mountaineering and skiing equipment, where its lightweight and robust characteristics make it ideal for items such as mountaineering sticks, spikes, ski poles, and ice skates. Titanium's use extends to fencing gear, fishing equipment, rowing parts, and track and field athletics spikes, highlighting its versatility across a diverse range of sporting goods [67, 68].

3.4.2 Watches and jewelry

Titanium-based alloys are increasingly popular in watchmaking and jewelry due to their durability, corrosion resistance, and hypoallergenic properties. Watches made from titanium alloys are lightweight, scratch-resistant, and highly robust. In jewelry, titanium alloys offer a modern esthetic, durability, and resistance to tarnishing, appealing to consumers seeking high-quality and long-lasting pieces [69].

3.4.3 Eyewear and fashion accessories

In the realm of fashion accessories, titanium-based alloys find application in eyewear frames and fashion accessories like wallets, belt buckles, and phone cases. The strength and lightweight nature of these alloys contribute to comfortable and durable products, appealing to consumers looking for stylish yet functional accessories [70].

3.4.4 Consumer goods

Titanium-based alloys are also employed in various consumer goods, including camping equipment, cookware, and electronic devices. The corrosion resistance, heat resistance, and lightweight properties of these alloys contribute to durable and high-performance consumer products [71].

Titanium-based alloys have found diverse applications in sports equipment and consumer goods, offering a balance of strength, durability, and lightweight properties. Their use in sports gear, watches, jewelry, fashion accessories, and various consumer products continues to drive innovation and cater to consumer demands for high-performance and long-lasting products [67].

3.5 Applications of titanium-based alloys in construction

Titanium-based alloys have emerged as innovative materials in the construction industry due to their unique properties, offering durability, corrosion resistance, and strength. This chapter explores the diverse applications of titanium alloys in construction, focusing on architectural structures, infrastructure, and specialty applications.

3.5.1 Architectural structures

Titanium-based alloys find application in architectural structures due to their esthetic appeal, durability, and corrosion resistance. The lightweight nature of these alloys allows for innovative and unique designs in buildings, facades, and artistic installations [72].

3.5.2 Infrastructure

In infrastructure projects, titanium-based alloys are utilized in bridge components, reinforcing bars, and cladding due to their corrosion resistance and long-term durability. These alloys contribute to extending the lifespan of structures exposed to harsh environmental conditions [73].

3.5.3 Specialty applications

Titanium alloys are used in specialty construction applications such as roofing materials, high-performance coatings, and seismic reinforcement due to their strength, lightness, and corrosion resistance. These applications contribute to enhanced structural integrity and longevity [74].

3.5.4 Sustainable construction

The use of titanium-based alloys in construction aligns with sustainability goals due to their recyclability, longevity, and resistance to corrosion. Their contribution to reducing maintenance and replacement needs aligns with sustainable construction practices [75].

Titanium-based alloys offer unique advantages in the construction industry, contributing to innovative architectural designs, resilient infrastructure, and specialty applications. Their properties of strength, corrosion resistance, and sustainability continue to drive their adoption in various construction projects [72].

4. Conclusion

Titanium-based alloys offer a wide range of applications across industries, driven by their unique combination of properties such as strength, lightweight, corrosion resistance, and biocompatibility. This chapter has provided a thorough examination of titanium-based alloys, showcasing their classification and extensive applications.

The chapter contains collected and organized information from the most recent studies. By categorizing these alloys based on their metallurgical structure and alloying elements into three main groups (alpha, beta, and alpha-beta alloys) and two subcategories (near alpha and meta-stable beta alloys), engineers and researchers have unlocked a multitude of possibilities for their utilization across diverse industries. From aerospace to medical, automotive, sports, and construction, titanium alloys continue to revolutionize various sectors, contributing to advancements in technology, healthcare, and infrastructure. As research and development efforts continue to advance, titanium-based alloys are expected to play an increasingly significant role in shaping the future of engineering and materials science. With ongoing optimization of their properties and exploration of new applications, these alloys hold immense promise for driving innovation and revolutionizing industries worldwide.

Author details

Nada H.A. Besisa* and Takeaki Yajima
Department of Electrical and Electronic Engineering, Kyushu University, Fukuoka, Japan

*Address all correspondence to: nadaamin129@gmail.com

References

[1] Pushp P et al. Classification and applications of titanium and its alloys. Materials Today: Proceedings. 2022;**54**(2):537-542

[2] Veiga C et al. Properties and applications of titanium alloys: A brief review. Reviews on Advanced Materials Science. 2012;**32**:133-148

[3] Liang Y, Wang HY. Influence of prior-β-grain size on tensile strength of a laser-deposited α/β titanium alloy at room and elevated temperatures. Materials Science and Engineering A. 2015;**622**:16-20

[4] Ye X, Lingsheng W, Song G. Retraction notice to effects of high-energy electro-pulsing treatment on microstructure, mechanical properties and corrosion behavior of Ti–6Al–4V alloy. Materials Science and Engineering: C. 2015;**49**:851-860

[5] Peters M et al. In: Leyens C, Peters M, editors. Titanium and Titanium Alloys. Germany: Wiley-VCH; 2003. p. 1

[6] Joshi VA. Titanium Alloys: An Atlas of Structures and Fracture Features. 6000 Broken Sound Parkway NW, Suite 300 Boca Raton, FL: CRC Press-Taylor & Francis Group; 2006. pp. 33487-32742

[7] Lütjering G, Williams JC. Titanium. Germany: Springer Verlag; 2007

[8] Sibum H. In: Leyens C, Peters M, editors. Titanium and Titanium Alloys. Germany: Wiley-VCH; 2003. p. 231

[9] Marin E et al. Biomedical applications of titanium alloys: A comprehensive review. Materials. 2024;**17**:114

[10] Mwinteribo TV et al. Mechanical properties of near alpha titanium alloys for high-temperature applications-a review. Aircraft Engineering and Aerospace Technology. 2020;**92**:521-540

[11] Sidhu SS et al. A review on alloy design, biological response, and strengthening of β-titanium alloys as biomaterials. Materials Science & Engineering. C, Materials for Biological Applications. 2021;**121**:111661

[12] Liu Y et al. Unravelling the competitive effect of microstructural features on the fracture toughness and tensile properties of near beta titanium alloys. Journal of Materials Science and Technology. 2022;**97**:101-112

[13] Alipour S et al. Nitinol: From historical milestones to functional properties and biomedical applications. Proceedings of the Institution of Mechanical Engineers. Part H. 2022;**236**:1595-1612

[14] Donachie MJ. Titanium: A Technical Guide. Metals Park, OH 44073: ASM International; 1988

[15] Williams JC. Titanium alloys for aerospace applications. Titanium. 2003;**2**:843-898

[16] Boyer R. An overview on the use of titanium in the aerospace industry. Materials Science and Engineering: A. 1996;**213**:103-114

[17] Wang L et al. Recent advances in titanium alloys for aerospace applications. Materials Science and Engineering: A. 2020;**770**:138625

[18] Froes FH et al. The role of titanium alloying additions in the development of high-strength, high-temperature aerospace alloys. JOM. 1994;**46**(9):30-35

[19] Chai GB et al. Application of titanium alloys in aerospace industry. Advanced Materials Research. 2013;**781-784**:803-806

[20] Banerjee D et al. Titanium alloys in satellite and spacecraft: Properties and applications. Progress in Aerospace Sciences. 2011;**47**(5):400-413

[21] Lupinc F et al. Titanium alloys in aerospace application. Aircraft Engineering and Aerospace Technology. 2000;**72**(1):73-79

[22] Long M, Rack HJ. Titanium alloys in total joint replacement- a materials science perspective. Biomaterials. 1998;**19**(18):1621-1639

[23] Niinomi M. Mechanical biocompatibilities of titanium alloys for biomedical applications. Journal of the Mechanical Behavior of Biomedical Materials. 2002;**1**(1):30-42

[24] Mahmud A et al. Mechanical behavior assessment of ti-6al-4v ELI alloy produced by laser powder bed fusion. Metals. 2021;**11**:1671

[25] Brantley WA. Evolution, clinical applications, and prospects of nickel-titanium alloys for orthodontic purposes. Journal of the World Federation of Orthodontists. 2020;**9**:S19-S26

[26] Madhukar S. Review on use of titanium and its alloys as implants in dental applications. International Journal of Current Engineering and Technology. 2020;**10**:513-517

[27] Sales PHDH et al. Do zirconia dental implants present better clinical results than titanium dental implants? A systematic review and meta-analysis. Journal of Stomatology Oral and Maxillofacial Surgery. 2023;**124**:101324

[28] Campanelli LC. A review on the recent advances concerning the fatigue performance of titanium alloys for orthopedic applications. Journal of Materials Research. 2021;**36**:151-165

[29] Aufa AN et al. Recent advances in ti-6al-4v additively manufactured by selective laser melting for biomedical implants: Prospect development. Journal of Alloys and Compounds. 2022;**896**:163072

[30] Mohanta M et al. Commercial pure titanium-a potential candidate for cardiovascular stent. Materwiss. Werksttech. 2022;**53**:1518-1543

[31] Nagaraja S et al. Oxide layer formation, corrosion, and biocompatibility of nitinol cardiovascular devices. Shape Memory and Superelasticity. 2022;**8**:45-63

[32] Ponsonnet L et al. Surface treatments of titanium alloy implants for the enhancement of osseointegration. Journal of Biomedical Materials Research. 2000;**51**(2):302-311

[33] Norlin A et al. Investigation of Pt, Ti, TiN, and nano-porous carbon electrodes for implantable cardioverter-defibrillator applications. Electrochimica Acta. 2004;**49**:4011-4020

[34] Beshchasna N et al. Recent advances in manufacturing innovative stents. Pharmaceutics. 2020;**12**:349

[35] Bhuvaneshwar GS et al. Evaluation of materials for artificial heart valves. Bulletin of Materials Science. 1991;**14**:1363-1374

[36] Olszyna A, Smolik J. Nanocrystalline diamond-like carbon coatings produced on the N4–TiC composites intended for the edges of cutting tools. Thin Solid Films. 2004;**459**:224-227

[37] Castagnini F et al. Highly porous titanium cup in cementless total hip arthroplasty: Registry results at eight years. International Orthopaedics. 2019;**43**:1815-1821

[38] De Meo F et al. Trabecular titanium acetabular cups in hip revision surgery: Mid-term clinical and radiological outcomes. Hip International. 2018;**28**:61-65

[39] Hell AK et al. The vertical expandable prosthetic titanium rib implant for the treatment of thoracic insufficiency syndrome associated with congenital and neuromuscular scoliosis in young children. Journal of Pediatric Orthopaedics. Part B. 2005;**14**:287-293

[40] Zuo W et al. Properties improvement of titanium alloys scaffolds in bone tissue engineering: A literature review. Annals of Translational Medicine. 2021;**9**:1259

[41] Geetha M et al. Biodegradable magnesium for orthopedic applications: A review on corrosion, biocompatibility, and surface modifications. Materials Science and Engineering C. 2014;**43**:287-299

[42] Block MS. Dental implants: The last 100 years. Journal of Oral and Maxillofacial Surgery. 2018;**76**:11-26

[43] Lambjerg-Hansen H, Asmussen E. Mechanical properties of endodontic posts. Journal of Oral Rehabilitation. 2008;**24**:882-887

[44] Khan MA et al. Titanium and its alloys as potential materials for medical devices. Advances in Materials Science and Engineering. 2015;**2015**:1-14

[45] Bombač D et al. Materials & Geoenvironment. 2007;**54**:471

[46] Elias CN et al. JOM journal of the minerals. Metals and Materials Society. 2008;**60**:46

[47] Banovetz JM et al. Titanium plate fixation: A review of implant failures. Journal of Orthopaedic Trauma. 1996;**10**:389-394

[48] Perren SM et al. How to choose between the implant materials steel and titanium in orthopedic trauma surgery: Part 2-biological aspects. Acta Chirurgiae Orthopaedicae et Traumatologiae Čechoslovaca. 2017;**84**:85-90

[49] Arens S et al. Influence of materials for fixation implants on local infection. An experimental study of steel versus titanium DCP in rabbits. Journal of Bone & Joint Surgery. British Volume. 1996;**78**:647-651

[50] Mohi Eldin MM, Ali AMA. Lumbar transpedicular implant failure: A clinical and surgical challenge and its radiological assessment. Asian the Spine Journal. 2014;**8**:281-297

[51] Szypryt P, Forward D. The use and abuse of locking plates. Orthopaedic Trauma. 2009;**23**:281-290

[52] Jost B et al. Locking plate fixation of fractures of the proximal humerus: Analysis of complications, revision strategies and outcome. Journal of Shoulder and Elbow Surgery. 2013;**22**:542-549

[53] He G, Hagiwara M. Titanium and Titanium Alloy Applications in Medicine. Materials Science and Engineering: C. 2006;**26**(14):533-576

[54] Niinomi M et al. Materials Science and Engineering: C. 2005;**25**:417

[55] Hao YL et al. Young's modulus and mechanical properties of Ti-29Nb-13Ta-4.6Zr in relation to α martensite. Metallurgical and Materials Transactions A. 2002;**33**:3137-3144

[56] Long M et al. Physical and mechanical properties of titanium alloys: A review. Proceedings of the Institution of Mechanical Engineers, Part L: Journal of Materials: Design and Applications. 2010;**224**(1):1-16

[57] Haase F et al. Two novel titanium alloys for medical applications: Thermo-mechanical treatment, mechanical properties, and fracture analysis. Journal of Materials Research. 2022;**37**:2589-2603

[58] Lütjering G, Williams JC. Titanium. Kyoto: Springer Science & Business Media; 2007

[59] Banerjee D, Williams JC. Perspectives on titanium science and technology. Acta Materialia. 2013;**61**(3):844-879

[60] Lee W-S et al. Titanium and titanium alloys as lightweight materials for automotive applications. Materials. 2014;7(1):642-650

[61] Lütjering G. Influence of processing on microstructure and mechanical properties of (α+β) titanium alloys. Materials Science and Engineering: A. 1998;**243**(1-2):32-45

[62] Banerjee D et al. Overview: Titanium in marine applications. Materials Science and Engineering: A. 1998;**243**(1-2):206-214

[63] Liu Y et al. A review on recent developments for design, synthesis, and application of titanium-based alloys in automotive applications. Rare Metals. 2015;**34**(11):779-792

[64] Hideki F et al. Shinnittetsu Giho. 2003;**378**:62

[65] Gupta RK, Ojha SN. Recent developments on titanium and titanium alloys for biomedical applications. Materials Science and Engineering: C. 2012;**32**(4):1069-1078

[66] Froes FH, Qian M, Grove W. Titanium in consumer applications. Materials & Design. 1999;**20**(1):3-10

[67] Subramanian V et al. Titanium in sports. Advanced Materials Research. 2013;**825**:221-226

[68] Li M et al. Application and optimization design of titanium alloy in sports equipment. Journal of Physics: Conference Series. 2021;**1820**:012011

[69] Hahn T et al. Titanium and titanium alloys as jewellery material: Advantages and limitations. Key Engineering Materials. 2008;**373**:163-170

[70] Park B et al. The properties and applications of titanium in fashion industry. Textile Research Journal. 2016;**86**(13):1461-1474

[71] Zhou Y et al. Titanium and titanium alloys in consumer goods. Advanced Materials Research. 2014;**849**:325-330

[72] Cao X et al. The use of titanium in construction industry. Materials Science Forum. 2014;**849**:1079-1085

[73] Zhang Y et al. Titanium in civil engineering: A review. Advanced Materials Research. 2014;**891-892**:1016-1021

[74] Ahmad Z et al. Titanium alloys for architectural applications: A review. Key Engineering Materials. 2017;**758**:42-49

[75] Ashby MF et al. Material selection in sustainable construction. Materials & Design. 2010;**31**(1):599-606

Chapter 7

Mechanical and Microstructural Characterization of the Ti-6Al-4V Alloy Processed by Additive Manufacturing for Overdenture Prosthesis

Mariana Correa Rossi, Angel Vicente Escuder, Ruben Agustin Panadero, Miguel Gomez Pólo, Pedro Peñalver and Vicente Amigó Borrás

Abstract

The main objective of this work is to show the capabilities of additive manufacturing to obtain arches and overdentures from titanium alloys. Overdentures are obtained mainly by subtractive techniques in both titanium alloys and Co-Cr-Mo. Obtaining these overdentures in Ti-6Al-4V, with better biocompatibility than Co alloys, by additive manufacturing (AM), by both laser and electron beam techniques, is of increasing interest. However, adequate mechanical and microstructural characterization is necessary to bring them closer to the alloys obtained by forging and machining. Parts obtained by selective laser melting (SLM) have been developed, which show mechanical properties like those of casting and plastic deformation, although their plasticity decreases significantly. Its lamellar microstructure can be modified by thermal treatments that improve the plasticity of AM alloys, which currently present a deformation slightly lower than that required by the American Society for Testing and Materials (ASTM) F2924-2021 standard. Therefore, there is a need to improve this property through appropriate thermal treatments. Its lamellar microstructure can be modified through heat treatments that can improve the plasticity of MA alloys, which currently have a deformation slightly lower than that required by the ASTM F2924-2021 standard. Hence, there is a need to improve this property through thermal treatments.

Keywords: overdenture prosthesis, additive manufacturing, selective laser melting, microstructure, small punch, mechanical properties, EBSD

IntechOpen

1. Introduction

Additive manufacturing (AM) or 3D printing, is a recent technique that makes it possible to produce low-cost materials in a short time, using computer-aided design CAD, 3D Slash, 3DPrinterOS, FreeCAD, Fusion 360, or TinkerCAD prototypes that can be transformed into 3D objects, which are built layer by layer [1, 2] or scans of a real object, based on the tissue of the host [3]. In this field, there are several technologies capable of using these complex materials in engineering, medicine, aerospace, automotive, construction, and dentistry.

The American Society for Testing and Materials (ASTM) defines AM as the "process of joining materials to make parts from 3D model data, usually layer upon layer" [4]. More specifically, in the biomedical area, AM techniques have been used to create solid or porous metallic implants, through different methods of AM. The conventional manufacturing methods present some disadvantages in terms of machining, standard sizes and make it difficult to obtain porous structures that facilitate osseointegration process, which is the mechanical stability of the metal within the bone tissue.

The complex parts must efficiently serve as excellent biocompatibility, biofunctionality, good hardness, low elastic modulus, low susceptibility to corrosion, wear resistance, and fatigue strength [5]. One of their main applications has been the manufacture of personalized medical implants using computerized tomography scans of a patient's affected region [6]. This device is quite difficult to obtain through conventional methods such as computer numerical control (CNC) machining. This is why this technique is being widely used in different areas. During fabrication, it includes various near-net shaping routes capable of manufacturing 3D geometries directly from raw materials, demanding little follow-on post-processing [7].

In addition to powders, wire-type materials can also be used as feedstock materials, and wire-based AM methods such as wire arc additive manufacturing are gaining significant attention in the additive manufacturing of ($\alpha + \beta$) Ti6Al4V alloys [8]. The common feature of these AM processes is the use of geometrical data, which is sliced into layers with a defined thickness depending on the application. Following the sliced pattern, a focused, high-power laser or electron beam scans and melts the pioneer powders, producing a molten pool. As the heating source moves away, the molten pool cools down faster and solidifies to form a track bead. This procedure is continual until a final geometry is completed.

Newly, AM technique uses titanium alloys to produce custom-made implants for hip joints and dental prosthesis [9, 10], which is the subject of this chapter. The main AM technique used nowadays to create complex geometric implants is powder bed fusion (PBF). The PBF technique employs a high-energy source that selectively melts a bed of powdered material (preferably metallic materials) layer by layer. The PBF technique can be separated into two procedures depending on the type of energy source used [11]. **Figure 1** indicates a schematic procedure of the PBF technique, which is formed by two fabrication routes: electron beam melting (EBM) and selective beam melting (SLM).

The first one is called electron beam melting (EBM), which uses an electron beam, **Figure 2**. The EBM was proposed and commercialized first by Arcam AB in 2001 [12]. The EBM is also a powder bed fusion process, but its heating source is an electron beam instead of a laser beam. Because of the special working nature of electron beam, EBM builds parts in a high-vacuum environment of 10^{-4} mbar or greater, providing an ideal contamination-free environment for manufacturing reactive materials (such as titanium) that have a high affinity to oxygen and nitrogen. Additionally,

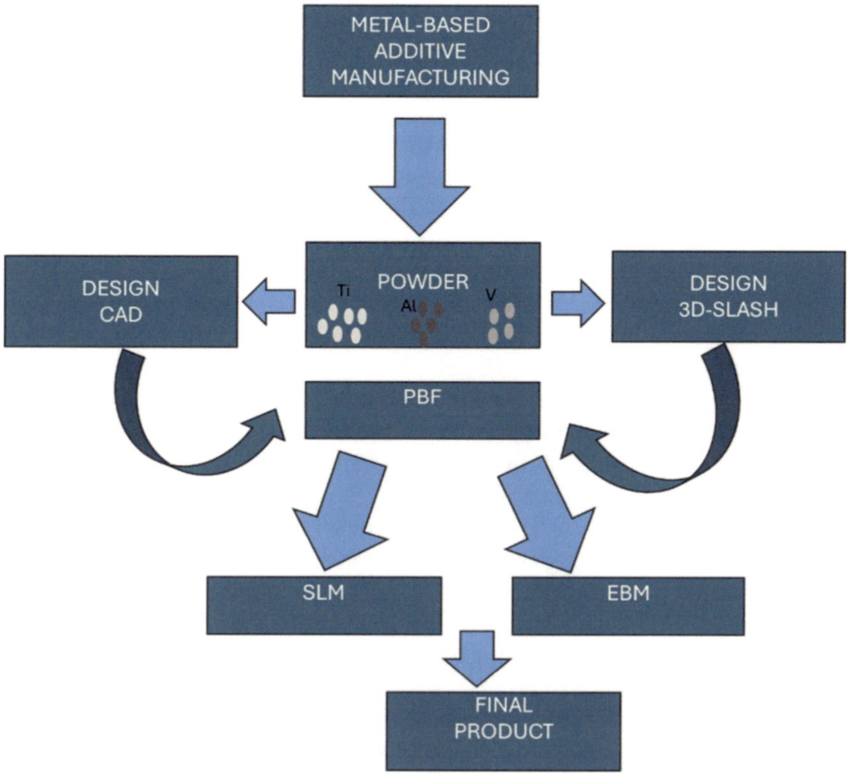

Figure 1.
Schematic procedures of selective laser melting and electron beam melting used for fabrication of complex parts by powder bed fusion technique.

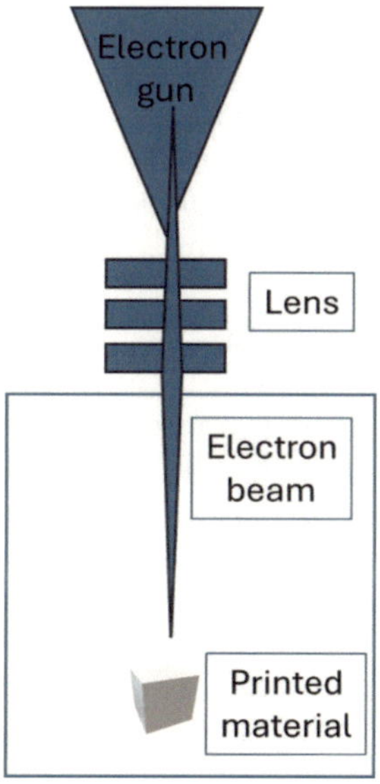

Figure 2.
Schematic procedures of electron beam melting used for fabrication of complex parts.

this technique can generate a faster build rate due to its superior energy input and fast scan speed [13]. The high build temperature of 600–750°C also leads to different manufacturing features.

Besides, the EBM method offers the advantage of manufacturing prostheses with open-cell structures (lattice structures) to obtain properties like those of a bone (e.g., low elastic modulus). For instance, some works have been showed that EBM technology can produce porous of (α + β) Ti4Al6V alloy with a compression strength and a Young's modulus compatible with those of a natural bone (1.03–17.5 GPa) [14–17]. This technique uses a high-intensity energy source (a fiber laser). To fabricate the parts, a layer of material powder with a thickness varying between 0 and 100 mm is spread over the table. Then, the recoating blade distributes the powder across the build table that was preheated.

Subsequently, the laser selectively melts the powder layer based on the CAD model [18]. The parts produced by means of this technique are built in two steps. First, the contour of the part is built, and then, the powder inside the contour is melted until the complete part is obtained [19]. For the EBM process, the temperature in the molten pool was estimated to reach 2700°C [20]. Price et al. [21] conducted a specific study using thermography and measured a temperature of 2500°C in the molten pool. Al-Bermani et al. [22] calculated the cooling rate of an EBM-fabricated Ti6Al4V bead on a stainless-steel substrate and obtained a cooling rate between 103 and 105 K/s. Antonysamy et al. [23] also obtained a cooling rate in the order of 10^4 s through simulation.

The second technique that PBF includes is selective laser melting (SLM), **Figure 3**, which employs a high-power laser [19]. It involves melting a powder bed using an electron beam as the energy source, and the whole process is carried out in a vacuum chamber [24]. Arcam AB, a Swiss brand founded in 1997 and later acquired by GE Additive in 2017, is currently the leader in EBM prototyping, with both companies being pioneers in this technique [12]. The SLM process started in 1995 at the Fraunhofer Institute ILT [25]. In SLM, metallic powders are uniformly spread on the building platform by rake instead of being blown out from nozzles. A focused laser beam scans the surface according to the prescribed path and selectively melts the powders in this layer, after which a new layer of powders is spread after lowering the building platform to the distance of the layer thickness. The layer height of SLM is on a scale of tens of microns, which is much thinner than that of EBM products [26]. The non-melted powders are left in the powder bed to support the subsequent layers. Powders surrounding the deposited parts are affected by the thermal process and cannot be reused due to the change in physical properties.

Besides being employed to manufacture customized dental implants [27], complex biomedical structures [28], and porous parts (e.g., bone substitutes) [29], this method has shown promising results in hip arthroplasty (particularly resurfacing hip arthroplasty [RHA]) [30]. However, in laser technology, materials melt while the parts are at relatively low temperatures, which may induce residual stresses that can cause the components in service to fail. To reduce the effects of excessive stress, some important factors to consider include implant's rigidity, as well as implant–bone fit and adhesion.

Like SLM, the components fabricated using EBM are also surrounded by partly melted metallic powders, and thus, conduction dominates the heat transfer in the EBM process. Heat loss via radiation also plays a role but can be neglected as compared to conduction. Due to the highly concentrated energy source and extremely short interaction time, a high temperature and cooling rate will be produced in the molten pool.

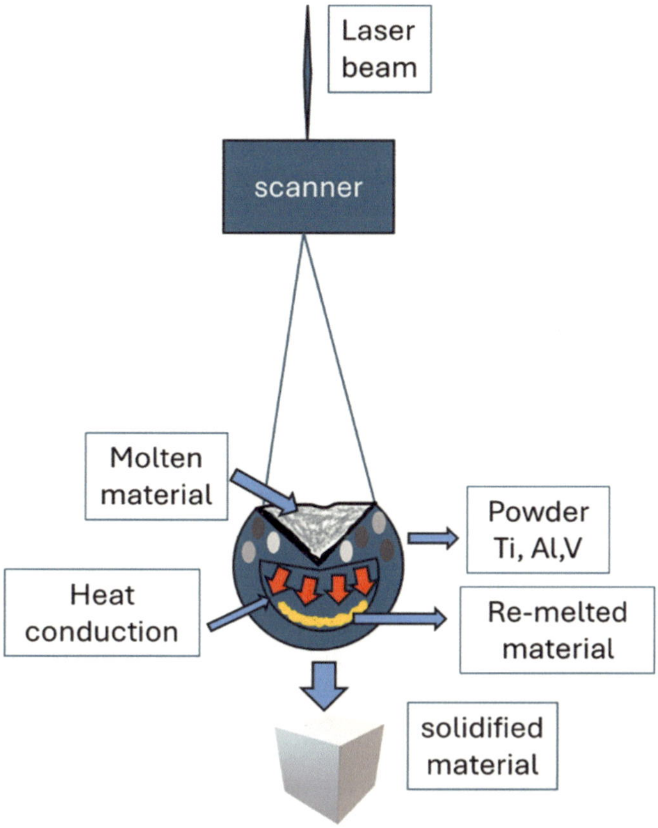

Figure 3.
Schematic procedures of selective laser melting used for fabrication of complex parts.

Yadroitsev et al. [31] reported a maximum temperature of about 2710 K in the Ti6Al4V molten pool created by SLM, which also produces a cooling rate in the range of about 104–10^6 K/s [32]. To precisely understand the thermal behavior of AM processes, the authors' group developed a laser direct deposition model to analyze the temperature history for a three-track laser directly deposited Ti6Al4V. The model considers laser-powder interaction, mass addition, heat transfer, and fluid dynamics in the molten pool occurring in the laser direct deposition process.

Actual deposition parameters and temperature-dependent material properties were used as inputs in the model. The extracted free surface, molten pool, and heat-affected zone boundaries that are superposed on the experimental micrograph, which show an excellent agreement. The temporal and spatial temperature fields in the molten pool generated at 11 different locations along the dashed line on the micrograph are also shown. A maximum temperature of approximately 2600 K is generated in the molten pool, and it decreases to room temperature within about 0.25 s as the laser beam moves away.

The average cooling rate in the molten pool is about 10^4 K/s, and even in the heat-affected zone, an average cooling rate of about 5×10^3 K/s is still obtained. As the laser scan speed for this deposition is 15 mm/s, even for scanning a small layer of 1 cm^2, the temperature has dropped to near room temperature before the laser travels back to the same location for the subsequent layer. Stevens et al. [33] have studied binder jetting of Ti-6Al-4V alloy and showed that the sintered density at the edges and regions with significant topological curvature is lower and higher than in the other areas, respectively. The microstructural heterogeneity is ascribed to the powder-binder interactions during printing. Dilip et al. [34] have studied the feasibility of TiAl fabrication by binder jetting of Ti-6Al-4V/Al powder mixture followed by reactive sintering. They have demonstrated the formation of TiAl with various other intermediate phases.

Titanium and its alloys are used in biomedical implants due to their excellent biological performance, such as biocompatibility (non-toxic and low allergenic properties) and osseointegration [35]. It is pertinent to point out that the Ti system is complex; hence, the properties of Ti alloys greatly vary with the chemical composition, level of impurities, phase constitution, grain size, etc. [36]. Nevertheless, demands for orthopedic implants (artificial knees, hip joints, elbows, bone plates, and screws for fracture fixation), as well as dental implants (removable prostheses, maxillofacial prostheses, and supporting materials), are increasing.

According to ISO 5832 standards, 26 groups of titanium alloys have been used in biomedical applications; among them, Ti6Al4V (grade 5) is the most used [2]. This grade is heat treatable with decent combination of properties and fabricability. The remarkable biocompatibility of titanium alloys makes them ideal candidates to replace hard tissues and implant them in the body. This chapter will describe and discuss the Ti64 Gd23 fabricated by powder by fusion (PBF), as well as the influence of the process on the mechanical and electrochemical properties.

2. A brief description of Ti-6Al-4V powders and its process of fabrication by PBD and its characterization

The powder used was LPW Ti64 Gd23, supplied by LPW Technology Ltd., in accordance with the ASTM B348/B348M-19 standard [37]. **Figure 4A,B** indicate the morphology of the Ti64 powder and its granulometry, which presents many of it of size 40 μm with a collection of tiny satellites and some agglomerate (**Figure 4A**), as can be seen in the size distribution of **Figure 4B**. In this granulometric distribution, a small peak is observed corresponding to a small volume fraction of about 10 μm, resulting in values of d(0.1) = 25.6 μm, d(0.5) = 38.1 μm, and d(0.9) = 103.2 μm.

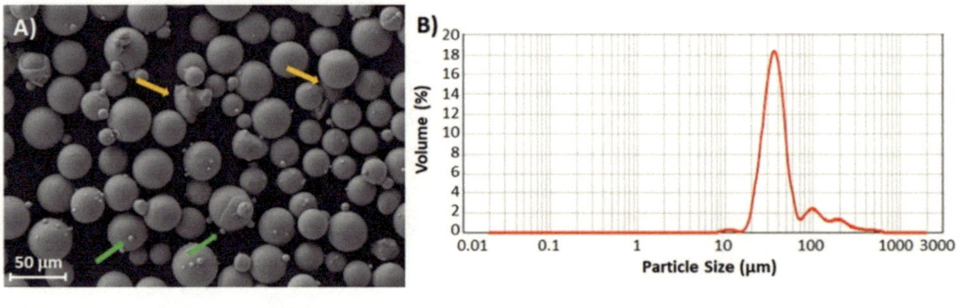

Figure 4.
(A) Secondary electron (SE) image of the powder used in printing components (green arrows indicate small satellites and yellow arrows irregular melts) and (B) granulometric distribution of the powder.

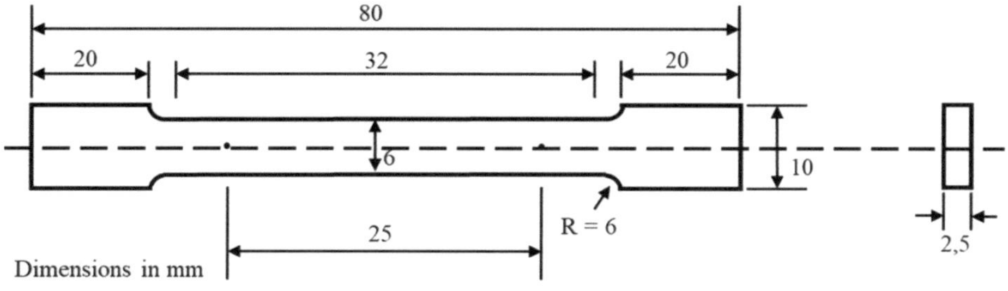

Figure 5.
Dimensions of the tensile specimen obtained in the X/Y and Z axes.

The test specimens have been obtained by the PBF process using a Trumpf TruPrint 1000 equipment from TRUMPF Laser- und Systemtechnik GmbH, operating with 200 W TRUMPF fiber lasers (wavelength: 1.070 nm), in an argon atmosphere and a scanning speed of 70 mm/s. The thickness of the layers has been 20 μm. Tensile specimens have been obtained in the Ti-6Al-4V ELI (Extra Low Interstitial) with PBF. The tensile tests are carried out in accordance with ASTM E8/E8M–22 [38], with the dimensions indicated in **Figure 5**. The compression tests have been carried out in accordance with the ASTM E9–18 standard [39] in cylinders obtained in the Z axis with a diameter of 13 mm and heights of 25 mm, with a ratio of 2.0 and 38 mm for a ratio of 3.0.

The structural analysis has been carried out using the Bruker D2Phaser X-ray equipment, using copper Kα radiation (1.5406 Å) at 30 kV and 15 mA, at a 2θ angle between 20° and 90°, with a step of 0.02° and a scanning speed of 0.0025°s^{-1}. The analysis of the results was carried out using the Rietveld method with the Materials Analysis Using Diffraction (MAUD) software version 2.9993. The microstructure was studied, after its metallographic preparation, using a NIKON LV100 optical microscope and a Zeiss Ultra 55 field emission scanning microscope (FE-SEM) equipped with X-ray spectroscopy (EDS) microanalysis from Oxford Instruments Ltda. To reveal the microstructure, it has been etching the surface polished using Kroll's reagent (HNO_3 6 mL, HF 3 mL, 100 mL distilled water).

Elastic modulus (E) was measured using the impulse excitation technique (IET, ATCP-Sonelastic®). The ATCP software Sonelastic 3.0 was used to analyze the data. Tensile test was performed on samples with a calibrated geometry following the ASTM standard. A Shimadzu universal testing machine model, Autograph GX Plus 100 kN was used with an optical extensometer, at a crosshead speed of 0.5 mm·min^{-1}. The compression tests were carried out on an Ibertest MEH 2000 kN universal testing machine. The hardness has been determined using a Centaur HD9-45 durometer on the HR15T superficial Rockwell scale, applying 147 N for 15 seconds. The microhardness scans were carried out in a Shimadzu HMV microhardness tester with an applied load of 2.94 N g and an application time of 15 s.

Corrosion tests were performed by potentiostat AUTOLAB PGSTAT204 with a three-electrode cell configuration (Ag/AgCl electrode, platinum wire, and the investigated alloy) at 37°C and a scanning rate [40]. The Ringer–Hartmann solution (NaCl 5.97 g/L, KCl 0.37 g/L, $CaCl_2$ 0.22 g/L, Na lactate 3.25 g/L, pH 6.5) was used as the electrolytes [41]. Electrochemical impedance spectroscopy (EIS) measurements, from 10 mHz to 100 kHz, were fitted by the ZView program (version 3.5a from Scribner Associates Inc., USA) for two-layer equivalent circuit simulation.

3. Physical, geometrical, and structural properties of powders and pieces obtained by PBF

The properties of the alloy obtained by additive manufacturing are comparable to those of the forging alloy, according to the ASTM F3001-14 (2021) standard [42]. The general density of the forging Ti-6Al-4V ELI alloy is 4.43 g cm^{-3}. The determination of the densities obtained in the analyzed samples is shown in **Table 1**.

As shown in the table, the 3D printed material has a slightly lower density than the wrought alloy, so it can finally incorporate up to 1.3% porosity. This can affect the mechanical properties, but above all, it can also affect its plasticity and fatigue resistance. This porosity is somewhat higher than that described by Dong et al. using powder modification through mechanical alloying and printing by selective laser melting (SLM) [43].

The X-ray diffraction is similar for both the powder and the forging material and the additive manufacturing material (see **Figure 6**). The diffraction angles are practically the same, although with a different intensity. While for the starting powder, the β-Ti phase is practically negligible, and the highest intensities occur in the (101) α-Ti plane, in additive manufacturing, the (200) β-Ti angle and the maximum intensity

Alloy and process	Open porosity (%)	Close porosity (%)	Archimedes density ($g{\cdot}cm^{-3}$)	Relative density (%)
Ti-6al-4V ELI forge	—	—	4.432 ± 0.008	
Ti-6Al-4V AM cylinder piece	0.20 ± 0.08	1.96 ± 0.41	4.412 ± 0.020	98.70 ± 0.44

Table 1.
Densities obtained in Ti-6Al-4V ELI alloys from forging and additive manufacturing samples.

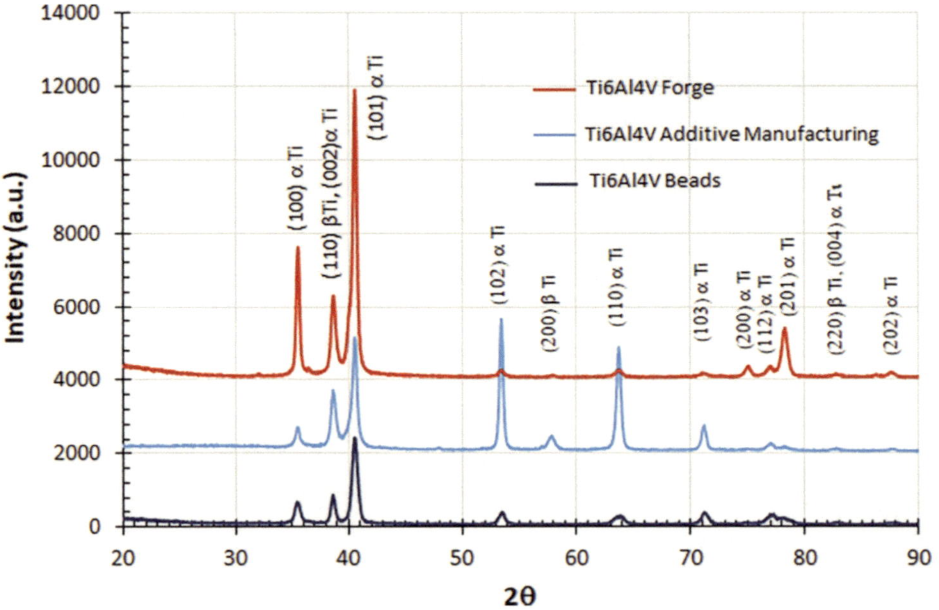

Figure 6.
X-ray diffraction patterns of Ti-6Al-4V ELI alloy from the powders, forging and additive manufacturing pieces.

peaks correspond to the α-Ti phase to the (101) (102) (110) planes. However, the forging material presents its maximum intensity in the plane (101) corresponding to the α-Ti phase, next to the plane (100) of the same phase. However, works such as those of Dong et al. [43] use SLM to observe the most intense peaks at (100) and (101), more like the peak intensity of wrought Ti-6Al-4V. However, this depends on the orientation of the laser during printing and the direction of growth of the layers. In our case, there are notable differences in the intensity of the peaks corresponding to each family of planes, which are more similar to the powder used in the process.

The hardness obtained in the pieces for the tensile test manufactured in the Z-direction has presented an average of 74.1 ± 5.8 HR15T, equivalent to 98.2 ± 8.0 HV. For the specimens manufactured in the X/Y direction, the hardness obtained was 79.1 ± 1.7 HR15T equivalent to 118.0 ± 6.9 HV. However, the hardness obtained in the first layers obtained was 51.7 HR15T (43.2 HV). This macroscopic hardness contrasts with the results of the microhardness scans, expressed on the Vickers (HV) scale, which are shown in **Table 2**, both in its radial and longitudinal directions. **Figure 7** presents the evolution of the measurements carried out along the diameter and longitudinally to the central axis of the specimens obtained in the direction of the Z axis. The average value obtained similarly in both axes is represented in solid lines, 353 ± 6 HV (longitudinal) and 347 ± 7 HV (radial), and very similar to that of the forge.

Process	Hardness HR15T	Microhardness HV	Elastic Moduli E (GPa)
Ti-6al-4V ELI forge	89 ± 1	358 ± 2 HV	110 ± 1
Ti-6Al-4V AM Z axis	74 ± 6	347 ± 7 HV	104 ± 2
Ti-6Al-4V AM X/Y axis	79 ± 2	353 ± 6 HV	90 ± 1

Table 2.
Hardness, microhardness, and elastic modulus for the forged Ti-6Al-4V alloy and that obtained by additive manufacturing.

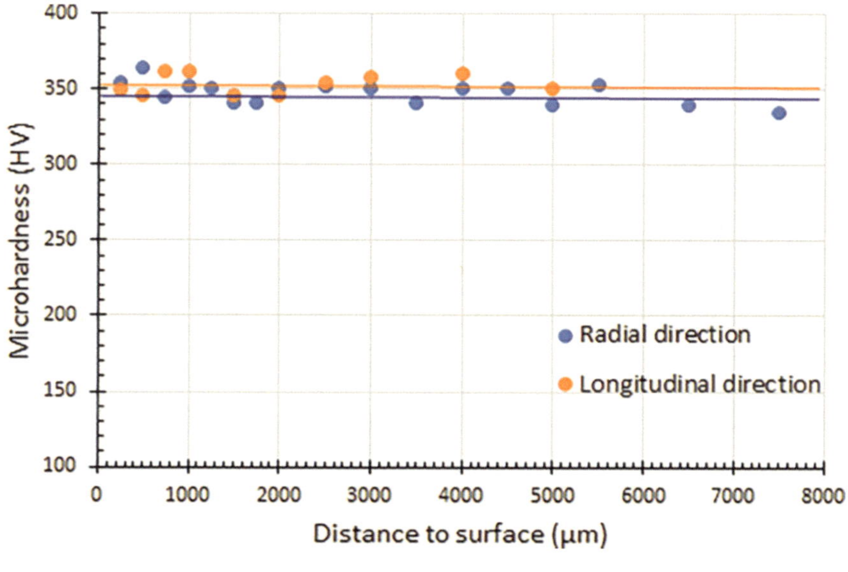

Figure 7.
Representation of the microhardness scans, both in the radial (X/Y axes) and longitudinal (Z axis) directions.

The elastic modulus obtained by the Sonelastic® method has been carried out both in the tensile samples, whose results are collected in **Table 2**, as in the cylindrical samples with a height of 38 mm, whose final average is 118 ± 2 GPa. There are real differences between the wrought alloy and the printed one with lower moduli, most likely due to the microstructural modification in the form of sheets and the porosity present. However, in the cylindrical pieces, there is a greater modulus that is very similar to that found during the tensile tests, as shown in **Table 3**. The results finally obtained depend greatly on the printing direction but also on the method of determining the elastic modulus, while in Sonelastic, the lowest value is obtained in the X/Y direction, with 90 GPa in the tensile test this elastic modulus increases to 117 GPa. However, in the Z-direction, both methods present values of 104 GPa, although in the case of the tensile test, the standard deviation is very high, approximately 27 GPa.

The tensile test was carried out in both the X/Y and Z positions. **Table 3** shows the results obtained in the X/Y position, in which the tensile elastic modulus is expressed in GPa, the limit elastic in MPa, the elastic deformation in %, the maximum resistance in MPa, and the maximum deformation in %. In this test, the deformation has been determined using a high-precision optical extensometer, so the final deformation corresponds perfectly to the real deformation.

Sample	Elastic modulus (GPa)	YS (MPa)	Elastic def. (%)	UTS (MPa)	Def. Max. (%)
Ti-6Al-4V AM (X/Y axis)	117 ± 9	942 ± 38	1 ± 0	1020 ± 46	1 ± 0
Ti-6Al-4V AM (Z axis)	104 ± 27	974 ± 27	1 ± 0	1100 ± 17	6 ± 1

Table 3.
Results obtained in the tensile tests in the manufacturing X/Y and Z positions.

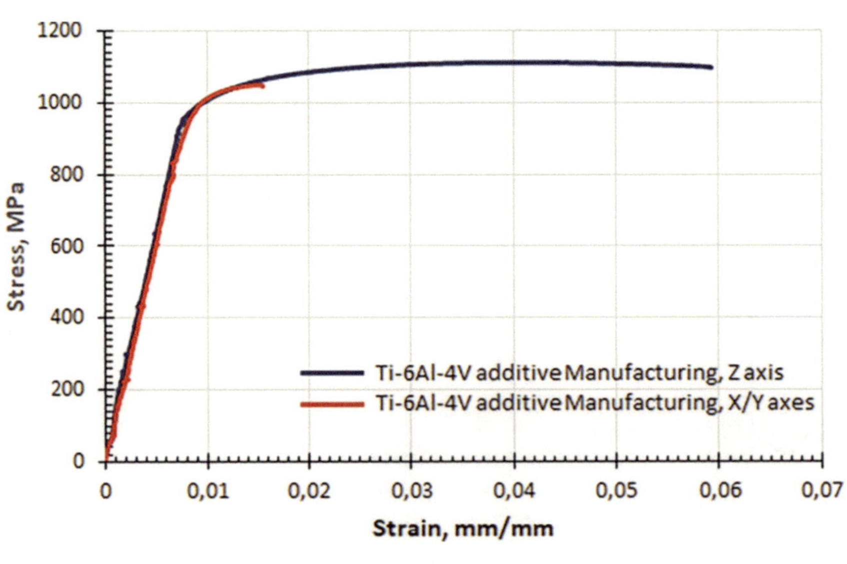

Figure 8.
Stress–strain diagram for the additive manufacturing samples in directions X/Y and Z.

As seen in **Table 3**, the mechanical properties are similar in the different directions, but the elasticity in this case is much higher, with deformation averages of 6% for position Z compared to 1% for position X/Y. **Figure 8** indicates the curves of stress–strain of the samples manufactured in both directions with a clear difference in the plastic behavior of the material. UTS is slightly lower than those obtained by Rafi et al. [44] of 1219 ± 20 MPa or the 1246 ± 143 MPa obtained by Wysocki et al. [26] and a lower yield strength. The measurement of the elongation (%) at the end of the breakage of the specimens, although with less precision than extensometer, has provided comparable results in any case. For the samples processed in X/Y, the average was 3 ± 0.4%, while, for the samples processed in Z, the average was 7 ± 0.4%. Considering the characteristics required by the ASTM F3001-14 standard [43], for class F, it more than meets the resistant properties, but not the elongation in the Z-direction. However, the maximum deformation reported by Wysocki et al. [26] is 3.2 ± 0.5% for the X/Y direction and 1.4 ± 0.5 in the Z-direction. In our tests, optical extensometry yields a plastic strain in the Z direction of 6 ± 1% higher than that reported by Wysocki et al. and of the order of that indicated by Rafi et al. [44] of 4.9 ± 0.6% in the built vertical direction. Much of these strain values are collected by Ngyuyen et al. [45] in their review carried out on additive manufacturing of Ti-6Al-4V alloy.

This lack of plastic deformation is mainly due to the defects that can occur between the laser fusion layers, as can be seen in **Figures 9A,B** and **10A,B**, in which a lack of fusion is observed in some areas.

The lack of plastic deformation is mainly due to defects that can occur between the layers fused by the laser. In **Figure 11A,B**, some of the most defined trajectories of the superposition of the fusion layers are represented with a dashed line, although they do not correspond to each one of them as they are much thinner.

The results obtained in the compression tests of the additive manufacturing samples, with a D/H ratio of 3.0, present a yield strength of 838 ± 85 MPa, with an elastic deformation of 1.95 ± 0.07%, somewhat higher than those obtained with traction. However, it has not been possible to determine the ultimate compressive strength (UTS), but the stress when deformation was 0.03 mm/mm, corresponding to a displacement of 1.14 mm of 953 ± 54 MPa. For the samples with D/H ratio = 2.0, a yield strength of 925 MPa, an elastic deformation of 2.4%, and a stress when the deformation was 3%, corresponding to a displacement of 0.75 mm of 1004 MPa, have been obtained.

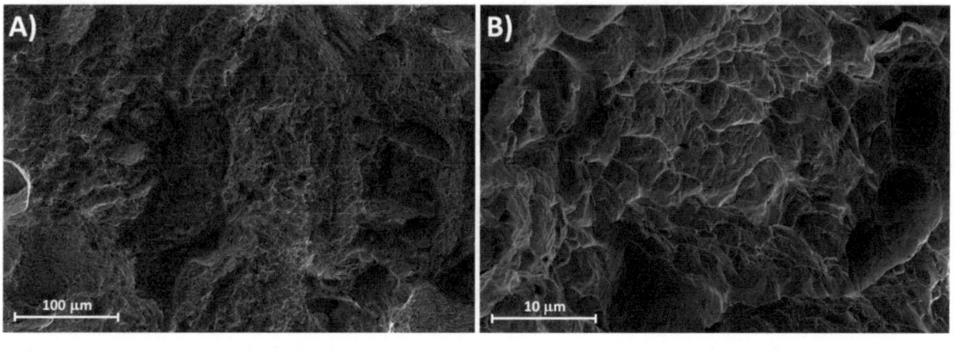

Figure 9.
Fractographies of the tensile fracture of the sample obtained by additive manufacturing in the X/Y direction. (A) General appearance of the fracture. (B) Detail of ductility with the formation of dimples.

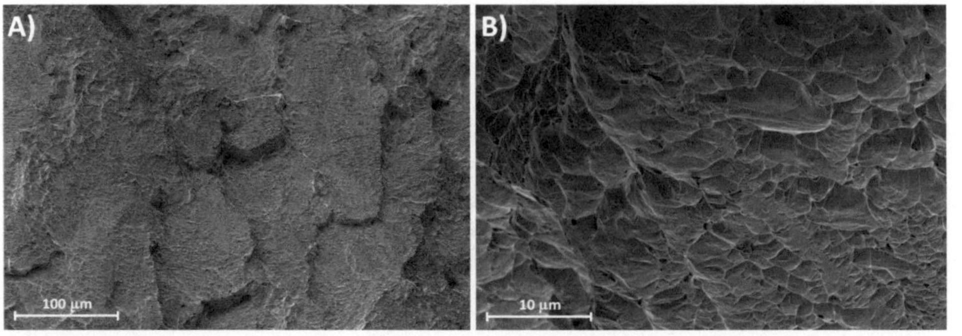

Figure 10.
Fractographies of the tensile fracture of the sample obtained by additive manufacturing in the Z-direction. (A) General appearance of the fracture. (B) Detail of the formation of ductile failure dimples.

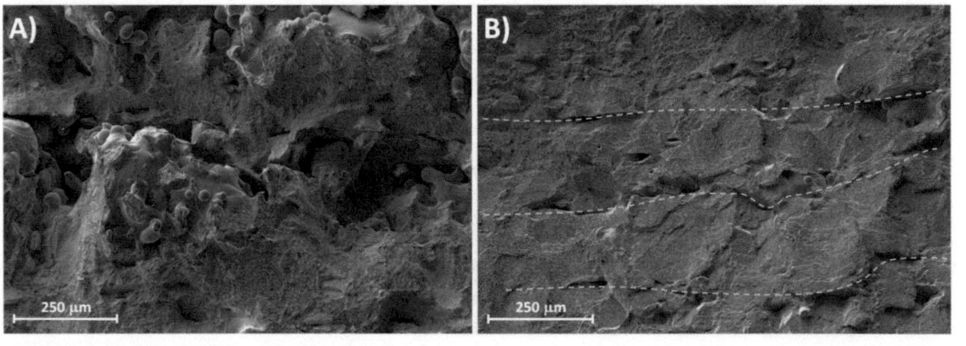

Figure 11.
(A) Infused material and cracking between layers in the X/Y direction. (B) Indications of crack paths between fusion layers in the X/Y direction.

The microstructure observed by optical microscopy, after etching with the Kroll solution, is shown in **Figure 12**. For the samples manufactured in the X/Y direction, in which some important defects originating from the fusion of the layers are observed. Mainly infused particles and small pores located between the fusion layers. **Figure 13** shows the microstructure of the printed samples in the Z-direction. This clearly shows the formation of the different fusion layers and how the lamellar microstructure originates through the layers, creating large grains corresponding to the β phase generated at high temperature.

However, the microstructure is much better appreciated after immersion corrosion tests. **Figure 14A–D** shows an important degradation located mainly in the α-Ti phase, especially when the magnification is higher. The microstructure of the formed alloy is mainly composed of equiaxed α phase grains with randomly positioned fine β phase grains. In the additive manufacturing samples, the formation of lamellar α + β grains is observed, with a strong presence of α phase at the grain boundary. This different distribution of the phases is what differentiates the behavior of the alloy depending on its production process. These microstructural differences are also observed when their microstructure and crystalline orientation are analyzed by backscattered electron diffraction. **Figure 15A–D** shows the far scattered diffraction (FSD) image. **Figure 15A** represents the band contrast, showing the morphology of the grains and the phase contrast. The distribution of the phases is indicated in **Figure 15B**. The inverse pole

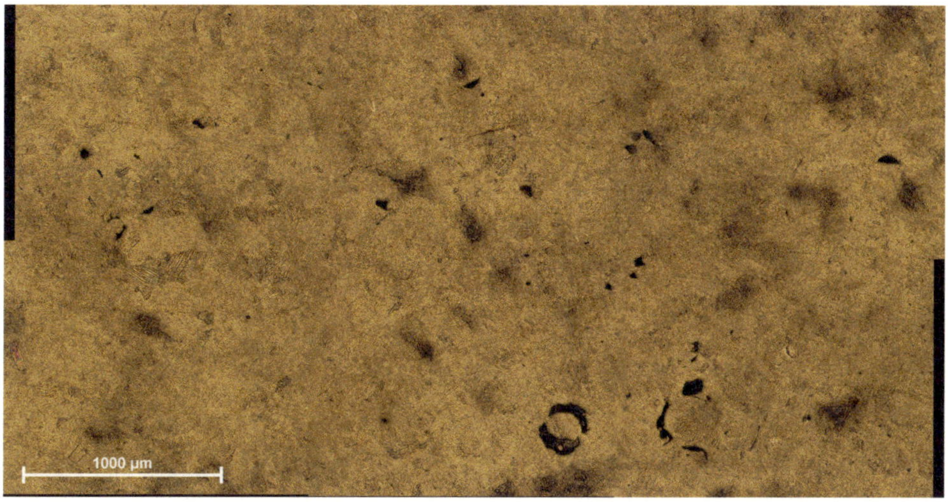

Figure 12.
Optical microscopy image of the microstructure of the Ti-6Al-4V alloy, obtained by additive manufacturing and printed in the X/Y direction.

Figure 13.
Optical microscopy image of the microstructure of the Ti-6Al-4V alloy, obtained by additive manufacturing printed in the Z-direction.

figure (IPF) is indicated in **Figure 15C**. In **Figure 15D** shows a slight texture in the (1010) plane. However, in AM Ti-6Al-4V, **Figure 16A–D**, at the magnification used, the size of the β phase grains obtained during solidification is not appreciated, and only the distribution of α phase sheets are appreciated, in which observe the formation of these sheets at angles of 60° or 120°, forming a Widmanstätten microstructure. Although a defined texture cannot be seen clearly, the orientation of these sheets bears no resemblance to the alloy obtained by forging. Therefore, the resistance and plasticity properties can be different between both types of alloys. This microstructure is

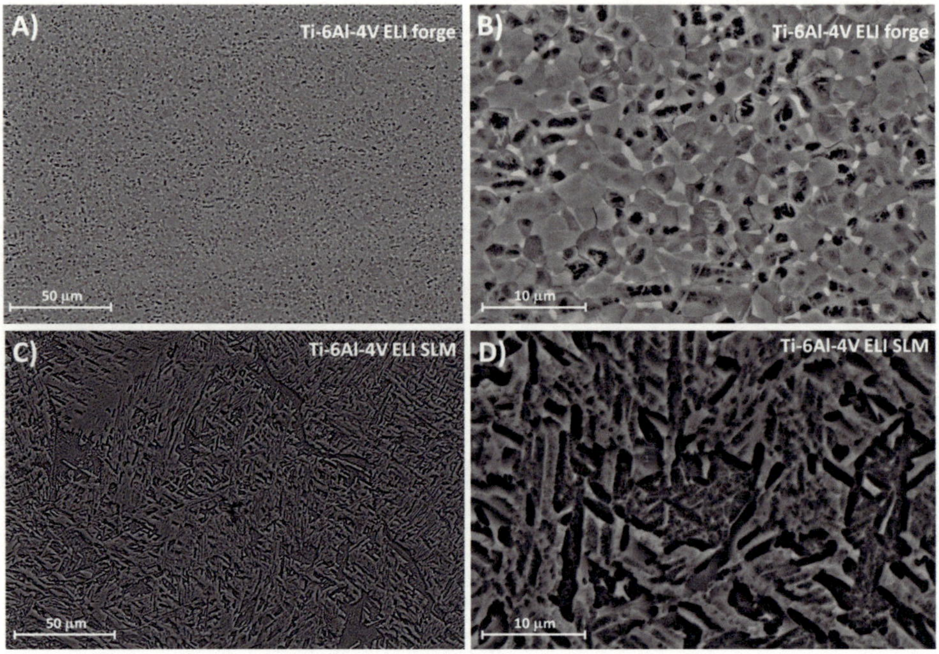

Figure 14.
Microstructure after immersion corrosion in Ringer-Hartmann electrolyte: (A) and (B) forged Ti-6Al-4V, (C) and (D) AM Ti-6Al-4V.

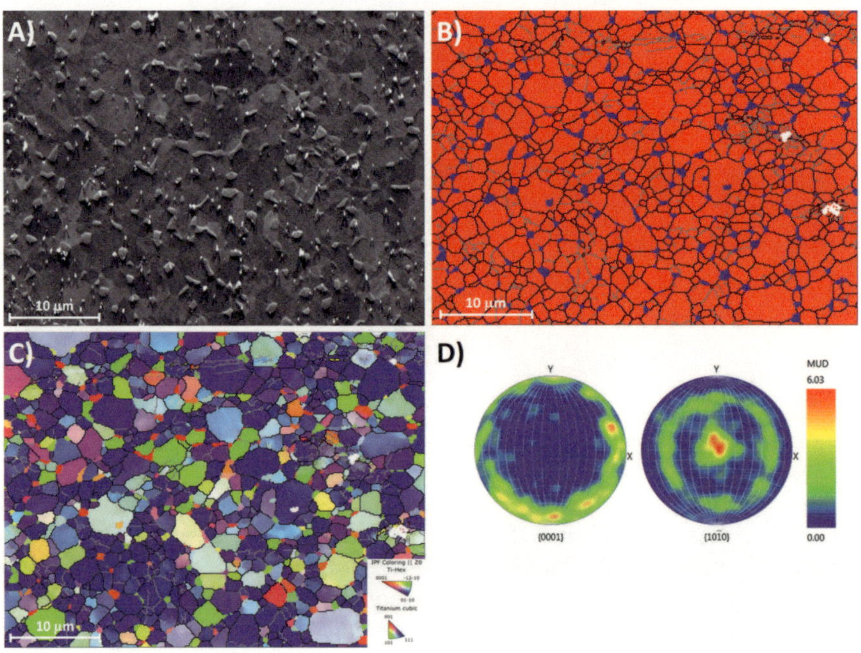

Figure 15.
Microstructure of forging Ti-6Al-4V ELI: (A) far scattered diffraction image. (B) Map of phase distribution (red represents the α-phase and the blue represents the β-phase). (C) Inverse pole figure map. (D) Pole figure in plans (0001) and (1010).

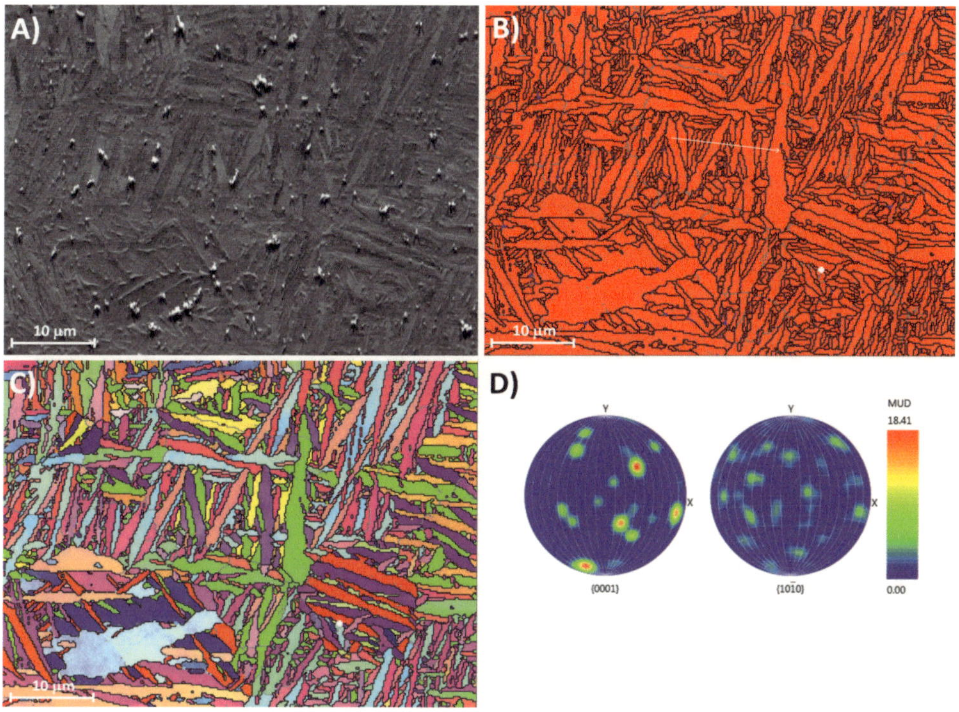

Figure 16.
Microstructure of AM Ti-6Al-4V ELI: (A) far scattered diffraction image. (B) Map of phase distribution. (C) Inverse pole figure map. (D) Pole figure in plans (0001) and (1010).

more like that obtained by EBM [26, 44] than that obtained by these same authors by SLM, whose main microstructure is α' martensite.

Corrosion resistance also presents differences between wrought and AM alloys. The potentiodynamic polarization curves, **Figure 17**, show great similarity between the CP-Ti samples and the wrought Ti-6Al-4V ElI alloy. However, the response of the additive manufacturing alloy, even presenting a more anodic potential and a lower passivation intensity than the two standard materials, shows a breakdown at a potential of 1.2 V and significantly increasing passivation intensity up to the end of the test. This could indicate that the rupture of the TiO_2 oxide layer initially formed. Although it could subsequently form again, limiting the current flow. The results of the corrosion parameters are shown in **Table 4**, which summarizes the open circuit potential (E_{OCP}), with the corrosion potential (E_{corr}), the current density (i_{corr}) obtained by the Tafel extrapolation method [46], the anodic (b_{an}) and cathodic (b_{cat}) Tafel constants, the polarization resistance (R_p), and finally the corrosion rate (C_r) expressed in µm·year^{-1}.

The Nyquist diagrams of the experiments carried out on Ti-6Al-4V AM and wrought Ti-6Al-4V ELI, together with the Ti CP, are shown in **Figure 18**. These diagrams have a semicircular arc shape, characteristic of passive metals, with important differences in them, finding the behavior of the AM alloy superior to CP-Ti but slightly inferior to wrought Ti-6Al-4V ELI alloy due to the greater length of the semicircle of this last alloy is associated with a greater resistance $R_{p.}$

In the Bode diagrams, **Figure 19**, a similar behavior also occurs with a phase angle close to 80°, which indicates a mainly capacitive behavior, although especially at low and medium frequencies. The impedance modulus is practically the same for the three

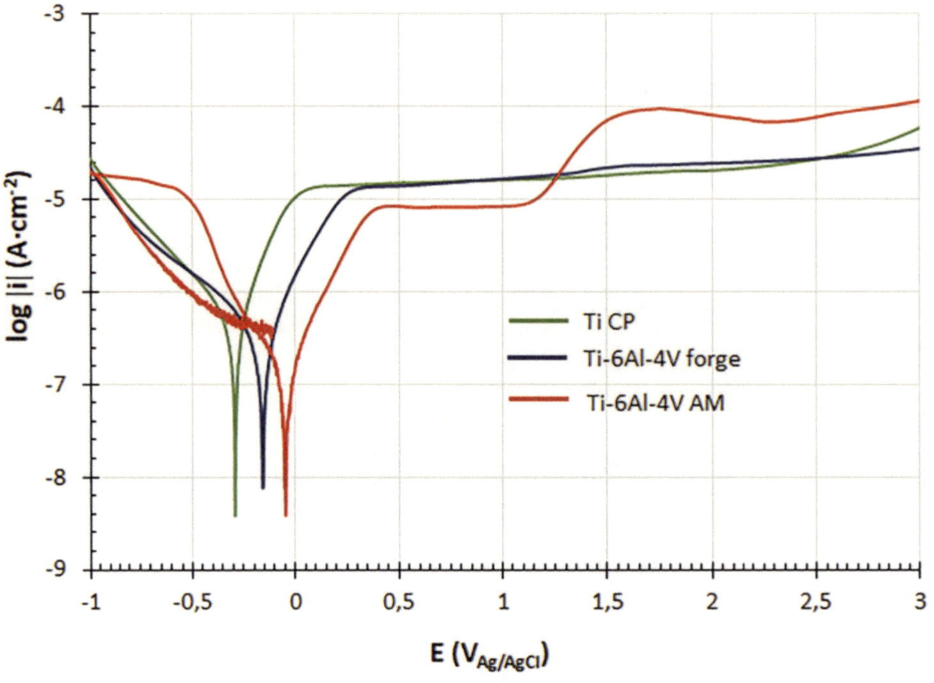

Figure 17.
Potentiodynamic polarization curves of the Ti-6Al-4V AM alloy in Ringer–Hartmann artificial saliva solution, and of CP-Ti and the wrought alloy.

Sample	E_{OCP} (V)	E_{corr} (V)	i_{corr} ($A \cdot cm^{-2}$)	b_{an}	b_{cat}	R_p ($k\Omega \cdot cm^2$)	C_r ($\mu m \cdot year^{-1}$)
Ti-6Al-4V AM	−0.08	−0.03	1.70×10^{-7}	0.12	0.12	361	1.68
CP-Ti	−0.24	−0.25	5.37×10^{-8}	0.10	0.11	109	4.59
wrought Ti-6Al-4V ELI	−0.02	−0.18	1.47×10^{-7}	0.11	0.12	231	1.25

Table 4.
Kinetics parameters obtained from the potentiodynamic polarization curves in the Ti-6Al-4V AM alloy, CP-Ti and the wrought Ti-6Al-4V ELI alloy.

materials, but the phase angle is slightly lower in Ti-6Al-4V AM, which would indicate a lower resistance, although it is quite like CP-Ti.

The results were compared with an equivalent double-layer circuit, using ZView software (version 3.5f), considering the non-ideal behavior with the electrolyte resistance (R_s), charge transfer resistance (R_{ct}), film resistance (R_{film}), and double-layer polarization resistance (CPE_{dl},) and film's (CPE_{film}). This circuit has been used by other authors, representing the equivalent circuit that best adapts to the composite corrosion mechanism [47]. In the same way, we can analyze the CPE exponents (n_{dl} and n_{film}) related to the capacities of the dense layer (C_{dl}) and the porous layer (C_{film}) [48]. These results are shown in **Table 5** for the three alloys analyzed, Ti-6Al-4V AM, CP-Ti, and wrought Ti-6Al-4V ELI.

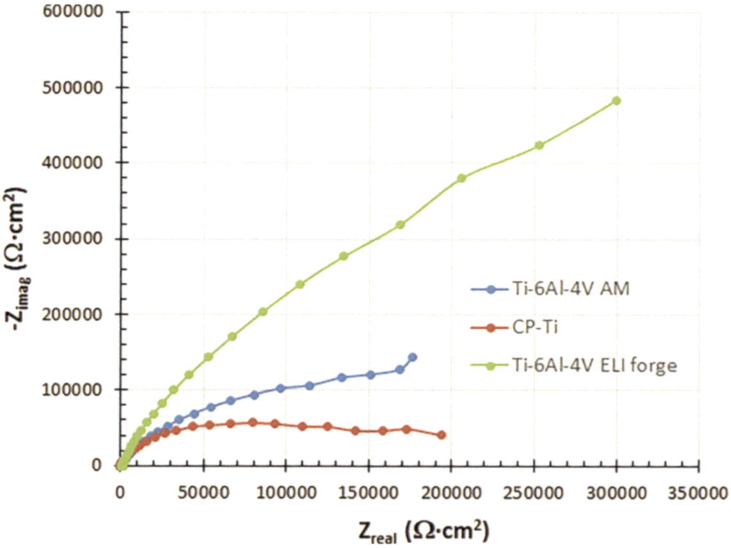

Figure 18.
Nyquist diagrams of the Ti-6Al-4V AM alloy in Ringer–Hartmann artificial saliva solution, and of CP-Ti and wrought Ti-6Al-4V ELI alloy.

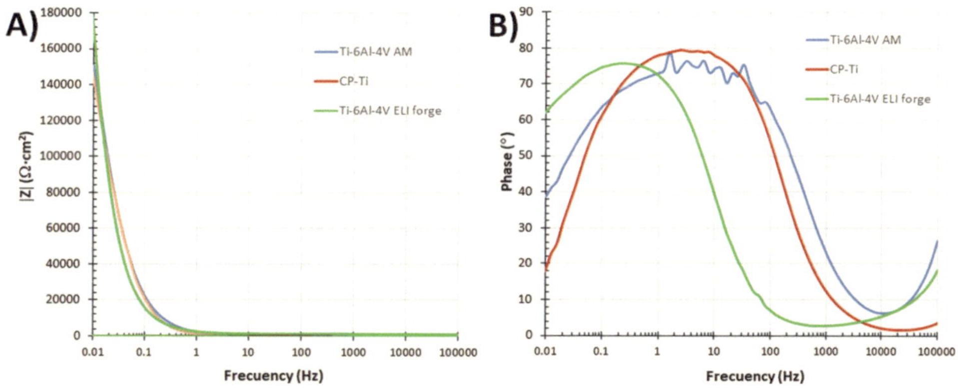

Figure 19.
Bode diagrams (modulus and phase) as a function of frequency for Ti-6Al-4V AM, CP-Ti, and wrought Ti-6Al-4V ELI alloy.

Sample	R_s ($\Omega \cdot cm^2$)	CPE_{dl} ($\mu F/cm^2$)	n_{dl}	R_{ct} ($k\Omega \cdot cm^2$)	CPE_{film} ($\mu F/cm^2$)	n_{film}	R_{film} ($k\Omega \cdot cm^2$)
Ti-6Al-4V AM	46	26	0.85	146	17	0.38	1100
CP-Ti	89	21	0.94	0.13	9	0.93	1380
Wrought Ti-6Al-4V	100	54	0.51	8.06	16	0.91	1190

Table 5.
Parameters obtained by electrochemical impedance spectrometry (EIS) in Ti-6Al-4V AM, Ti CP and wrought Ti-6Al-4V ELI alloys.

The values obtained for R_s are similar in all samples, also considering the obstacle of standardizing the resistivity of the electrolyte in the different tests. The charge transfer resistance is always lower than that of the film (R_{film}), which is related to the internal oxide layer as it is much more protective than the porous external layer [49]. However, the n_{dl} and n_{film} exponents were approximately 1 for CP-Ti. For the Ti-6Al-4V alloy, the n_{dl} exponent was 0.51 for the wrought alloy and the n_{film} of 0.38 for the AM alloy. Besides that, together with higher R_p, it ensures better behavior against corrosion compared to the CP-Ti. The fits obtained had a χ^2 in the order of 10^{-3}, which would indicate that the selected equivalent circuit is representative of the oxide layers formed in these alloys [49].

4. Discussion

Additive manufacturing offers excellent opportunities in the biomedical sector [50]. The selective laser melting (SLM) or direct metal laser sintering (DMLS) technology also offers great versatility and greater ease than other powder bed fusion (PBF) processes, such as electron beam melting (EBM) [51]. The obstacle of machining titanium alloys, as well as the residual stresses that they can accumulate, makes AM a perfectly applicable technology in different sectors mainly in the biomedical sector [52], where the customization of products is found. This is mostly justified, mainly in maxillofacial surgery or implant-supported prostheses [53]. The microstructure formed is basically laminar in practically all cases, with the formation of α phase at the grain boundary of solidification β phase. These β grains are large as they remain at high temperatures for some time in PBF processes [54]. From the grain boundary, the formation of the α phase begins in sheets with a Widmästaetten-type microstructure, **Figure 14**. However, the crystal structure is only modified by the cooling direction of the sample, which in our case enhances the (101) (102) and (110) planes, **Figure 6**.

The formation of the microstructure in PBF processes can originate, especially in laser-assisted processes, SLM or DMLS, metallurgical transformations to the martensite α'. Although this phase presents laminar α phase (HCP) microstructure, the mechanical properties obtained after the AM process are very similar to those presented by wrought Ti-6l-4V ELI alloy. However, the main difference is the ductility of the material, since it is decreased compared to the commercial forged alloy. However, it depends on the direction of the effort depending on the growth of the layers, **Table 3**. Similar results have been obtained in the works of Shunmugavel et al. [55], who, after a vacuum thermal treatment at 750°C obtained ductility properties that were also lower in the transverse direction, although in this case somewhat higher than those obtained after AM. In addition, a preferential formation of the primary β-phase can be available, as analyzed by Wei et al. using multi-laser and analyzing the mechanical properties of overlap regions. The transformation to α' phase is evident from the energy in the melting and cooling of the different layers [56]. The effect of energy in the process is analyzed by Cepeda-Jiménez et al. with a columnar formation of the β-phase prior to the α-α' transformation [57]. The application of a sub-β-transus heat treatment, the α phase is stabilized, with somewhat lower frequency at misorientation angles of 60°. Infusion defects are analyzed by Andani et al., who evaluate the spatter formation in the SLM process using multi-laser technology and conclude that most of the particle shape of the spatter is spherical and that this can significantly affect the fatigue properties of the components [58]. In fact, when determining their mechanical properties, a decreased elongation is present, especially when two-laser melting is used.

The microstructure conditions formed obviously affect the mechanical properties, which are also affected by external parameters such as surface roughness and possible internal defects in the material. These mechanical properties depend critically on the posterior heat treatment used. Although in all cases, the resistance decreases slightly, as the formation of the phase α phase stabilizes. At the same time, it is true that the longitudinal and transversal properties are getting closer together. Nevertheless, the biggest influence is on ductility, which decreases greatly when the load is applied perpendicular to the melting of the layers, i.e., in the Z-direction. Therefore, the stability of the α phase, simultaneously with the growth of the thickness of the α phase lamellae, promotes an increase in the ductility of the component [58], especially in the Z-direction. Rafi et al. [59] report only 5% elongations in both directions after heat treatment at 650°C for 4 hours Ar. Simonelli, Tse, and Tuck [60] obtain elongations higher than 10% with a treatment with N_2 at 730°C for 2 hours, while with an annealing treatment at 843°C for 2 hours and furnace cooling, Brandl, Leyens, and Palm [61] obtain elongations of 17% in the longitudinal direction and 15% in the transverse direction. However, it is true that in this case, the yield strength is 810–840 MPa and UTS between 860 and 920 MPa. Zhang et al. [11] obtain elongations higher than 20% when, after obtaining by L-PBF a phase distribution α' + (α + β), they subject the material to a sub-β-transus heat treatment at 950°C for 2 h, with subsequent cooling in air or in the furnace. However, one of the fundamental problems of AM is the lower reliability of the product due to the possible internal defects that can occur in the material. The differential melting at some points of the different layers usually generates irregular pores, **Figure 11**, or by microporosities derived from Ar absorption, although in this case, the pores are micrometric in size and rounded.

The elastic modulus obtained by impulse excitation technique in the cylindrical sheets is like that determined in the tensile tests in the X/Y direction, around 117 GPa, while in the Z-direction, it decreases to 104 GPa. The yield strength and maximum load exceed the minimum established by the ASTM standard, and the hardening index (YS/UTS) is 0.92 in the X/Y direction. For the Z-direction, use the YS/UTS. The microhardness does not present appreciable differences in both directions and with the wrought alloy, less than 2%.

However, the greatest difference is found in the plastic deformation where in the Z-direction, an elongation greater than 6% is obtained, as indicated in the ASTM F3001-14 standard. However, in the X/Y direction, a loss of 79% of the elongation with respect to Z is obtained, which is being far from the minimum indicated in the standard. This effect is seen in fractography where the influence on the fracture of the orientation of the layers and the influence of internal defects in the material are observed. Therefore, one of the biggest differences with wrought alloy is the greater difference in the results, which is expressed by high standard deviations, somewhat higher in the X/Y direction than in the Z-direction. Surface roughness also has an important effect on the resistant properties of the parts, which, jointly with possible internal defects, provide higher standard deviations than in wrought parts. This is one of the limitations of the technique in biomedical and especially aeronautical applications. A preliminary determination of the roughness obtained, using white light profilometry, has provided a R_a value of around 48 μm. This roughness and possible internal defects play a fundamental role in fatigue resistance and, hence, there is need to determine the number, shape, and size of these defects [62] and try to minimize them through thermal post-treatments or by hot isostatic pressing (HIP).

Thermal treatments, as already indicated, have a secondary effect on grain growth in the β phase and thickening of the microstructural lamellae in the α phase, but they can improve the ductility, which is necessary in these cases, by closing the internal pores, as gas porosity and other small fusion defects. Leuders et al. [63, 64] analyze the effect of ductility on the fatigue resistance of Ti-6Al-4V processed by SLM. In this sense, it is essential to evaluate the differences in microstructure in both static and dynamic properties. Greitemeier et al. [65] compare the fatigue resistance between electron beam melting (EBM) and direct metal laser sintering (DMLS), reducing the surface roughness, which possibly increases the fatigue resistance in both cases, with values close to 550 MPa for DMLS and 260 MPa for EBM, after 10^7 cycles.

The chemical and electrochemical behavior is also similar between the alloys obtained in both processes, wrought and AM. In immersion tests, small pits appear mainly in the α phase. On the other hand, due to the different equiaxial microstructure grains in one case and laminar in AM, the contact surface between the α and β phases is higher in AM, and therefore, the release of Ti ions is slightly higher. However, in its electrochemical behavior, the AM alloy has a lower E_{corr} than both the wrought alloy and CP-Ti, with great similarity in terms of i_{corr} and resistance to polarization R_p. Despite this, while wrought alloys present passivation potentials that extend beyond 3 V, the AM alloy presents a breakdown potential at 1.2 V, although it subsequently passivates again with higher passivation intensity values. In any case, a similar passive TiO_2 layer is developed in all cases that must be more widely studied. In any case, the potentiodynamic tests present variability in the three samples tested for each material. Thus, the value of the corrosion ratio is very similar for the two alloys Ti-6Al-4V ELI wrought and AM. In the electrochemical impedance spectrometry tests, it is observed that in the Niquist diagrams, the AM alloy is between the CP-Ti and the wrought alloy, but with similar behaviors. In the Bode diagrams, all the alloys show similar behaviors. The phase angles are around 80°, although in the case of the wrought alloy, it moves to low frequencies. With the results, an equivalent double-layer circuit has been obtained, the internal compact and the external porous, also used by other authors [44, 49, 66–68] with varying success. With the circuit and once the different resistive and capacitive components have been obtained, it is important to determine the CPE exponent, since an exponent greater than 0.8 confers a merely capacitive character and with values less than or close to 0.5 a diffusive character. As seen in the results table, very similar values are presented between the three alloys studied, and the AM alloy, it has a somewhat higher charge transfer resistance R_{ct}, although much lower than the resistance R_{film}, where the exponent n_{film} remains at 0.38.

5. Conclusions

The additively manufactured Ti-6Al-4V alloy has mechanical properties like the Ti-6Al-4V ELI wrought alloy. Its elastic modulus (104–90 GPa compared to 110 GPa for Ti-6Al-4V ELI), microhardness (347–353 HV compared to 358 HV for wrought alloy), yield strength (943–974 MPa), and maximum tensile strength (1020–1100 MPa) differ very little and, after some heat treatments, are even higher in the AM alloy.

Ductility, however, is limited and although the standards reduce the minimum ductility for AM alloys, these are not achieved when manufacturing is carried out in some specific directions, X/Y, since the maximum deformation is only 1.5%.

The microstructure is completely different, and the AM is very similar to that of casting, as it develops large β phase grains that, in their allotropic transformation like the Widmanstätten type α phase, but far from the microstructure of equiaxial grains in the wrought mill annealing alloy.

The X-ray diffraction indicates a distribution of similar phases in wrought and AM alloys, but with differences in textures that depend on the direction of growth of the layers but do not affect the mechanical properties except ductility.

The corrosion resistance is similar between AM and wrought Ti-6Al-4V alloy with very similar behavior and a similar corrosion ratio (1.68 $\mu m \cdot year^{-1}$ for AM alloy compared to 1.25 $\mu m \cdot year^{-1}$ for wrought alloy).

Therefore, no significant differences have been found in the behavior of AM alloys obtained from 3D printing, except for their microstructure that can hardly be modified, and the ductility derived from the manufacturing direction, internal defects, and surface roughness mainly. This makes the technique widely used in the manufacture of personalized parts in the biomedical sector, and possibly through subsequent heat treatments or hot isostatic processes, the reliability of the material is improved. In that case, its application in both overdentures and maxillofacial surgery, which require customization for each patient, can offer important advantages.

Author details

Mariana Correa Rossi[1], Angel Vicente Escuder[2], Ruben Agustin Panadero[3], Miguel Gomez Pólo[4], Pedro Peñalver[5] and Vicente Amigó Borrás[2]*

1 Materials Department, Universidade Federal de São Carlos (UFSCar), São Carlos, SP, Brazil

2 Universitat Politècnica de València, University Institute of Materials Technology, Valencia, Spain

3 Faculty of Medicine and Dentistry, Department of Stomatology, Universitat de València, Valencia, Spain

4 Faculty of Dentistry, Department of Conservative Dentistry and Oral Facial Prosthodontics, Universidad Complutense de Madrid, Spain

5 PromeDent Group, Madrid, Spain

*Address all correspondence to: vamigo@upv.edu.es

References

[1] Culmone C, Smit G. Additive manufacturing of medical instruments: A state-of-the-art review. Additive Manufacturing. 2019;**27**:461-473. DOI: 10.1016/j.addma.2019.03.015

[2] Tamayo JA, Riascos M, Vargas CA, Baena LM. Additive manufacturing of Ti6Al4V alloy via electron beam melting for the development of implants for the biomedical industry. Heliyon. 2021;**7**:e06892. DOI: 10.1016/j.heliyon.2021.e06892

[3] Kuang X et al. Self-enhancing sono-inks enable deep-penetration acoustic volumetric printing. Science. 2023;**382**:1148-1155. DOI: 10.1126/science.adi1563

[4] Bourell D, Kruth JP, Leu M, Levy G, Rosen D, Beese AM, et al. Materials for additive manufacturing. CIRP Annals. 2017;**66**:659-681. DOI: 10.1016/j.cirp.2017.05.009

[5] Breme H, Biehl V, Reger N, Gawalt E. Metallic biomaterials: Titanium and titanium alloys. In: Murphy GHW, Black J, editors. Handbook of Biomaterial Properties. New York: Springer; 2016. pp. 167-189

[6] Bartolo P, Kruth J-P, Silva J, Levy G, Malshe A, Rajurkar K, et al. Biomedical production of implants by additive electro-chemical and physical processes. CIRP Annals. 2012;**2012**(61):635-655. DOI: 10.1016/j.cirp.2012.05.005

[7] Imchi A, Petzoldt F, Hartwig T, et al. Microstructural development during additive manufacturing of biomedical grade Ti-6Al-4V alloy by three-dimensional binder jetting: Material aspects and mechanical properties. The International Journal of Advanced Manufacturing Technology. 2023;**127**:1541-1558. DOI: 10.1007/s00170-023-11661-1

[8] Bermingham MJ, Nicastro L, Kent D, Chen Y, Dargusch MS. Optimising the mechanical properties of Ti-6Al-4V components produced by wire + arc additive manufacturing with post-process heat treatments. Journal of Alloys and Compounds. 2018;**753**:247-255. DOI: 10.1016/j.jallcom.2018.04.158

[9] Yang F, Chen C, Zhou Q, et al. Laser beam melting 3D printing of Ti6Al4V based porous structured dental implants: Fabrication, biocompatibility analysis and photoelastic study. Scientific Reports. 2017;**7**:45360. DOI: 10.1038/srep45360

[10] Crenn MJ, Benoit A, Rohman G, Guilbert T, Chaussain C, Fromentin O, et al. Additive manufactured titanium for prosthetic application in dentistry: Surface topography characterization and in vitro cellular response of human gingival fibroblasts (HGFs). Computer Methods in Biomechanics and Biomedical Engineering. 2019;**22**:31-32. DOI: 10.1080/10255842.2020.1713467

[11] Zhang Y, Wu L, Guo X, Kane S, Deng Y, Jung Y-G, et al. Additive manufacturing of metallic materials: A review. JMEPEG. 2018;**27**:1-13. DOI: 10.1007/s11665-017-2747-y

[12] Andersson LE, Larsson M. Device and Arrangement for Producing a Three-Dimensional Object. Patent WO 2001081031 A1; Application Number: SE-0100932-W; 2001

[13] Li P, Warner DH, Fatemi A, Phan N. Critical assessment of the fatigue performance of additively manufactured

Ti–6Al–4V and perspective for future research. International Journal of Fatigue. 2016;**85**:130-143. DOI: 10.1016/j.ijfatigue.2015.12.0

[14] Rho JY, Ashman RB, Turner CH. Young's modulus of trabecular and cortical bone material: Ultrasonic and microtensile measurements. Journal of Biomechanics. 1993;**26**:111-119. DOI: 10.1016/0021-9290(93)90042-D

[15] Ashman RB, Jae R. Young elastic modulus of trabecular bone material. Journal of Biomechanics. 1988;**21**:177-181. DOI: 10.1016/0021-9290(88)90167-4

[16] Turner CH, Rho J, Takano Y, Tsui TY, Pharr GM. The elastic properties of trabecular and cortical bone tissues are similar: Results from two microscopic measurement techniques. Journal of Biomechanics. 1999;**32**:437-441. DOI: 10.1016/S0021-9290(98)00177-8

[17] Nicholson PHF, Cheng XG, Lowet G, Boonen S, Davie MWJ, Dequeker J, et al. Structural and material mechanical properties of human vertebral cancellous bone. Medical Engineering & Physics. 1997;**19**:729-737. DOI: 10.1016/s1350-4533(97)00030-1

[18] Liao Z, Zhang Y, Chen J, Ouyang J, Chen H. Finite element simulation of selective laser melting based on the layer thickness-dependent shrinkage ratio model. Optics & Laser Technology. 2024;**174**:110565. DOI: 10.1016/j.optlastec.2024.110565

[19] Ratna D. Chapter 2 - properties and processing of thermoset resin. In: Ratna D, editor. Recent Advances and Applications of Thermoset Resins. 2nd ed. Amsterdam, Netherlands: Elsevier; 2022. pp. 173-292. DOI: 10.1016/B978-0-323-85664-5.00003-X, ISBN 9780323856645

[20] Cheng B, Price S, Lydon J, Cooper K, Chou K. On process temperature in powder bed electron beam additive manufacturing: Model development and validation. Journal of Manufacturing Science and Engineering. 2014;**136**:061018. DOI: 10.1115/1.4028484

[21] Price S, Lydon J, Cooper K, Chou K. Experimental temperature analysis of powder-based electron beam additive manufacturing, 24th. In: Annual International Solid Freeform Fabrication Symposium; an Additive Manufacturing Conference, Proceedings. Austin, TX: University of Texas; 2013. pp. 162-173

[22] Al-Bermani SS, Blackmore ML, Zhang W, Todd I. The origin of microstructural diversity, texture, and mechanical properties in electron beam melted Ti-6Al-4V. Metallurgical and Materials Transactions AA. 2010;**41**:3422-3434. DOI: 10.1007/s11661-010-0397-x

[23] Antonysamy AA, Meyer J, Prangnell PB. Effect of build geometry on the β-grain structure and texture in additive manufacture of Ti6Al4V by selective electron beam melting. Materials Characterization. 2013;**84**:153-168. DOI: 10.1016/j.matchar.2013.07.012

[24] Cooper N, Coles LA, Everton S, Maskery I, Campion RP, Madkhaly S, et al. Additively manufactured ultra-high vacuum chamber for portable quantum technologies. Additive Manufacturing. 2021;**40**:101898. DOI: 10.1016/j.addma.2021.101898

[25] Brandt M. The role of lasers in additive manufacturing. In: Brandt M, editor. Laser Additive Manufacturing. Sawston, UK: Woodhead Publishing Limited; 2017. pp. 1-18

[26] Wysocki B, Maj P, Sitek R, Buhagiar J, Kurzydłowski K, Święszkowski W. Laser

and electron beam additive manufacturing methods of fabricating titanium bone implants. Applied Sciences. 2017;7:657. DOI: 10.3390/app7070657

[27] Bibb R, Eggbeer D, Williams R. Rapid manufacture of removable partial denture frameworks. Proceedings of the Institution of Mechanical Engineers, Part C. 2006;**12**:95-99. DOI: 10.1243/095441105X9372

[28] Wehmöller M, Warnke P, Zilian C, Eufinger H. Implant design and production—A new approach by selective laser melting. International Congress Series. 2005;**1281**:690-695. DOI: 10.1016/j.ics.2005.03.155

[29] Pattanayak D et al. Bioactive Ti metal analogous to human cancellous bone: Fabrication by selective laser melting and chemical treatments. Acta Biomaterialia. 2011;**7**:1398-1406. DOI: 10.1016/j.actbio.2010.09.034

[30] Uklejewski R, Winiecki M, Rogala P, Mielniczuk J. Selective laser melted prototype of original minimally invasive resurfacing hip endoprosthesis. Rapid Prototyping Journal. 2011;**17**:76-85. DOI: 10.1108/13552541111098653

[31] Yadroitsev I, Krakhmalev P, Yadroitsava I. Selective laser melting of Ti6Al4V alloy for biomedical applications: Temperature monitoring and microstructural evolution. Journal of Alloys and Compounds. 2014;**583**:404-409. DOI: 10.1016/j.jallcom.2013.08.183

[32] Pegues JW, Shao S, Shamsaei N, Sanaei N, Fatemi A, Warner DH, et al. Fatigue of additive manufactured Ti-6Al-4V, part I: The effects of powder feedstock, manufacturing, and post-process conditions on the resulting microstructure and defects. International Journal of Fatigue. 2020;**132**:105358. DOI: 10.1016/j.ijfatigue.2019.105358

[33] Stevens E, Schloder S, Bono E, et al. Density variation in binder jetting 3D-printed and sintered Ti-6Al-4V. Additive Manufacturing. 2018;**22**:746-752. DOI: 10.1016/j. addma. 2018.06.017

[34] Dilip JJS, Miyanaji H, Lassell A, et al. A novel method to fabricate TiAl intermetallic alloy 3D parts using additive manufacturing. Defence Technology. 2017;**13**:72-76. DOI: 10.1016/j. dt. 2016.08.001

[35] Khorasani AM, Goldberg M, Doeven EH, Littlefair G. Titanium in biomedical applications—Properties and fabrication: A review. Journal of Biomaterials and Tissue Engineering. 2015;**5**:593-619. DOI: 10.1166/jbt.2015.1361

[36] Salvador CAF, Maia EL, Costa FH, et al. A compilation of experimental data on the mechanical properties and microstructural features of Ti-alloys. Scientific Data. 2022;**9**:188. DOI: 10.1038/s41597-022-01283-9

[37] ASTM B348/B348M-19. Standard Specification for Titanium and Titanium Alloy Bars and Billets. West Conshohocken, Pensilvania, USA: ASTM International; 2019

[38] ASTM E8/E8M-22 Standard Test Methods for Tension Testing of Metallic Materials. West Conshohocken, Pensilvania, USA: ASTM International

[39] ASTM E9 - 09(2018). Standard Test Methods of Compression Testing of Metallic Materials at Room Temperature. West Conshohocken, Pensilvania, USA: ASTM international; 2018

[40] Zambrano JC, Dalmau A, Amigó V, Navarro-Laboulais J, Pereira JC.

Electrochemical corrosion behavior and mechanical properties of Ti–Ag biomedical alloys obtained by two powder metallurgy processing routes. Journal of the Mechanical Behavior of Biomedical Materials. 2020;**112**:104063. DOI: 10.1016/j.jmbbm.2

[41] Du J-K, Chao C-Y, Chiu K-Y, Chang Y-H, Chen K-K, Wu J-H, et al. Antibacterial properties and corrosion resistance of the newly developed biomaterial, Ti–12Nb–1Ag alloy. Metals. 2017;7:566. DOI: 10.3390/met7120566

[42] ASTM F3001-14. Standard Specification for Additive Manufacturing Titanium-6 Aluminum-4 Vanadium ELI (Extra Low Interstitial) with Powder Bed Fusion. West Conshohocken, Pensilvania, USA: ASTM International; 2021

[43] Dong YP, Tang JC, Wang DW, Wang N, He ZD, Li J, et al. Additive manufacturing of pure Ti with superior mechanical performance, low cost, and biocompatibility for potential replacement of Ti-6Al-4V. Materials and Design. 2020;**196**:109142. DOI: 10.1016/j.matdes.2020.109142

[44] Rafi H, Nadimpalli K, Gong H, Starr T, Stucker B. Microstructures and mechanical properties of Ti6Al4V parts fabricated by selective laser melting and electron beam melting. Journal of Materials Engineering and Performance. 2013;**22**:248. DOI: 10.1007/s11665-013-0658-0

[45] Nguyen HD, Pramanik A, Basak AK, Dong Y, Prakash C, Debnath S, et al. A critical review on additive manufacturing of Ti-6Al-4V alloy: Microstructure and mechanical properties. Journal of Materials Research and Technology. 2022;**18**:4641-4661. DOI: 10.1016/j.jmrt.2022.04.055

[46] Landolt D. Corrosion and Surface Chemistry of Metals. New York: EPFL Press; 2007. DOI: 10.1201/9781439807880

[47] Bolat G, Mareci D, Chelariu R, Izquierdo J, Gonzoález S, Souto RM. Investigation of the electrochemical behaviour of TiMo alloys in simulated physiological solutions. Electrochimica Acta. 2013;**113**:470-480. DOI: 10.1016/j.electacta.2013.09.116

[48] Dalmau A, Pina VG, Devesa F, Amigó V, Muñoz AI. Influence of fabrication process on electrochemical and surface properties of Ti–6Al–4V alloy for medical applications. Electrochimica Acta. 2013;**95**:102-111. DOI: 10.1016/j.electacta.2013.01.155

[49] Dalmau A, Pina VG, Devesa F, Amigó V, Muñoz AI. Electrochemical behavior of near-beta titanium biomedical alloys in phosphate buffer saline solution. Materials Science and Engineering C. 2015;**48**:55-62. DOI: 10.1016/j.msec.2014.11.036

[50] Harun WSW, Kamariah MSIN, Muhamad N, Ghani SAC, Ahmadd F, Mohamed Z. A review of powder additive manufacturing processes for metallic biomaterials. Powder Technology. 2018;**327**:128-151. DOI: 10.1016/j.powtec.2017.12.058

[51] Tofail SAM, Koumoulos EP, Bandyopadhyay A, Bose S, O'Donoghue L, Costas Charitidis C. Additive manufacturing: Scientific and technological challenges, market uptake and opportunities. Materials Today. 2018;**21**:22-37. DOI: 10.1016/j.mattod.2017.07.001

[52] Ngo TD, Kashani A, Imbalzano G, Nguyen KTQ, Hui D. Additive manufacturing (3D printing): A review of materials, methods, applications and challenges. Composites Part

B: Engineering. 2018;**143**:172-196. DOI: 10.1016/j.compositesb.2018.02.012

[53] Barbin T, Velôso DV, Del Rio SL, Borges GA, Presotto AGC, Barão VAR, et al. 3D metal printing in dentistry: An *in vitro* biomechanical comparative study of two additive manufacturing technologies for full-arch implant-supported prostheses. Journal of the Mechanical Behavior of Biomedical Materials. 2020;**108**:103821. DOI: 10.1016/j.jmbbm.2020.103821

[54] Frazier WE. Metal additive manufacturing: A review. Journal of Materials Engineering and Performance. 2014;**23**:1917-1928. DOI: 10.1007/s11665-014-0958-z

[55] Shunmugavel M, Polishetty A, Littlefair G. Microstructure and mechanical properties of wrought and additive manufactured Ti-6Al-4V cylindrical bars. Procedia Technology. 2015;**20**:231-236. DOI: 10.1016/j.protcy.2015.07.037

[56] Wei K, Li F, Huang G, Liu M, Deng J, He C, et al. Multi-laser powder bed fusion of Ti–6Al–4V alloy: Defect, microstructure, and mechanical property of overlap region. Materials Science & Engineering A. 2021;**802**:140644. DOI: 10.1016/j.msea.2020.140644

[57] Cepeda-Jiméneza CM, Potenza F, Magalini E, Luchin V, Molinari A, Pérez-Prado MT. Effect of energy density on the microstructure and texture evolution of Ti-6Al-4V manufactured by laser powder bed fusion. Materials. Characterization. 2020;**163**:110238. DOI: 10.1016/j.matchar.2020.110238

[58] Andani T, Dehghani R, Karamooz-Ravari MR, Reza Mirzaeifar R, Ni J. Spatter formation in selective laser melting process using multi-laser technology Mohsen. Materials & Design. 2017;**131**:460-469. DOI: 10.1016/j.matdes.2017.06.040

[59] Rafi HK, Starr TL, Stucker BE. A comparison of the tensile, fatigue, and fracture behavior of Ti-6Al-4V and 15-5 PH stainless steel parts made by selective laser melting. The International Journal of Advanced Manufacturing Technology. 2013;**69**:1299-1309. DOI: 10.1007/s00170-013-5106-7

[60] Simonelli M, Tse YY, Tuck C. Effect of the build orientation on the mechanical properties and fracture modes of SLM Ti–6Al–4V. Materials Science and Engineering: A. 2014;**616**:1-11. DOI: 10.1016/j.msea.2014.07.086

[61] Brandl E, Leyens C, Palm F. Mechanical properties of additive manufactured Ti-6Al-4V using wire and powder based processes. IOP Conference Series: Materials Science and Engineering. 2011;**26**:012004

[62] Fatemi A, Molaei R, Sharifimehr S, Phan N, Shamsaei N. Multiaxial fatigue behavior of wrought and additive manufactured Ti-6Al-4V including surface finish effect. International Journal of Fatigue. 2017;**100**:347-366. DOI: 10.1016/j.ijfatigue.2017.03.044

[63] Leuders S, Thöne M, Riemer A, Niendorf T, Tröster T, Richard H, et al. On the mechanical behaviour of titanium alloy TiAl6V4 manufactured by selective laser melting: Fatigue resistance and crack growth performance. International Journal of Fatigue. 2013;**48**:300-307. DOI: 10.1016/j.ijfatigue.2012.11.011

[64] Leuders S, Lieneke T, Lammers S, Tröster T, Niendorf T. On the fatigue properties of metals manufactured by selective laser melting–the role of ductility. Journal of Materials Research, Supl. Focus Issue: The Materials Science

of Additive Manufacturing; Warrendale. 2014;**29**:1911-1919. DOI: 10.1557/jmr.2014.157

[65] Greitemeier D, Palm F, Syassen F, Melz T. Fatigue performance of additive manufactured TiAl6V4 using electron and laser beam melting. International Journal of Fatigue. 2017;**94**:211-217. DOI: 10.1016/j.ijfatigue.2016.05.001

[66] da Silva Vieira Marques I, VAR B, da Cruz NC, Yuan JC-C, Mesquita MF, Ricomini-Filho AP, et al. Electrochemical behavior of bioactive coatings on cp-Ti surface for dental application. Corrosion Science. 2015;**100**:133-146. DOI: 10.1016/j.corsci.2015.07.019

[67] Xie F, He Xueming Lv Y, Wu M, He Xinbo QX. Selective laser sintered porous Ti–(4-10)Mo alloys for biomedical applications: Structural characteristics, mechanical properties and corrosion behaviour. Corrosion Science. 2015;**95**:117-124. DOI: 10.1016/j.corsci.2015.03.005

[68] Ureña J, Tsipas S, Pinto AM, Toptan F, Gordo E, Jiménez-Morales A. Corrosion and tribocorrosion behaviour of β-type Ti-Nb and Ti-Mo surfaces designed by diffusion treatments for biomedical applications. Corrosion Science. 2018;**140**:51-60. DOI: 10.1016/j.corsci.2018.06.024

Chapter 8

Residual Stress Analysis of Laser Cladded Commercially Pure Grade Titanium Alloy Plates

Tankiso Lawrence Ngake and Kadephi Vuyolwethu Mjali

Abstract

The aim of this study was to determine the influence of laser power on the microstructure, hardness, and residual stresses of laser-cladded titanium alloy. Laser powers of 600, 800, and 1000 W were varied while keeping other processing parameters constant. Microhardness and microstructure measurements were conducted using a Vickers microhardness tester and an optical microscope, respectively. Residual stresses were determined using an X-ray diffractometer (XRD). The results obtained revealed a non-linear relationship between microhardness and increasing laser power. Microstructural analysis indicated a transformation from all α grains to α' martensite in the cladded zone, observed across all laser powers. Residual stress measurements showed a consistent tensile trend, decreasing with increasing laser power. The issue of increasing tensile residual stress can be mitigated by adjusting the laser power with minimal impact on the microstructure. These findings highlight the significant influence of processing parameters on surface properties, suggesting the potential for optimizing laser cladding parameters to enhance material corrosion and abrasion resistance properties.

Keywords: titanium, laser cladding, residual stress, microstructure, microhardness

1. Introduction

Titanium (Ti) and its alloys are widely used in the biomedical, automotive, and aerospace industries [1, 2]. Their extensive use is mostly due to their superior mechanical and biocompatibility qualities. Ti exists in multiple crystallographic forms. At room temperature, Ti exists in the form of a hexagonal close-packed (hcp) crystal structure, also referred to as the "alpha" (α) phase. At 883°C, the Ti crystal structure transforms to body-centered cubic (bcc), which is called the "beta" (β) phase. Alloying elements are generally classified as α or β stabilizers. Alpha stabilizers increase the temperature at which the α phase is stable, whereas the beta stabilizers stabilize the β phase at lower temperatures. Aluminum and oxygen are examples of α-stabilizers, while common β-stabilizers include vanadium and molybdenum. Ti alloys are classified into three groups: α, β, and $\alpha + \beta$ alloys. Alpha alloys consist of all-α microstructure. A beta alloy is one in which the β phase does not undergo martensitic breakdown when a small amount of the material is quenched into freezing water above its β transus. The α + β alloys are made up of a mixture of α and β phases, which are stable at room temperature [3].

The microstructure of Ti alloys is strongly influenced by the processing history and surface treatment [4]. Commercially pure (CP) Ti is a single-phase (α) material, with its properties are controlled by chemistry (interstitial impurity elements) and grain size. CP Ti is divided into four grades according to its strength and permitted levels of impurity elements (iron, carbon, nitrogen, and oxygen). CP grade 2 Ti is widely used in chemical industries due of its outstanding corrosion resistance and good strength [5]. However, there is a need to enhance the properties of CP grade 2 Ti to broaden its applications in the aerospace industry. The material property requirements in the aerospace industry include good strength for fatigue life, which ultimately reduces the likelihood of cracking during high-speed cyclic operations. The presence of tensile residual stresses (TRS) during operations is the primary cause of cracks, as they initiate cracks on the material's surface [6].

Researchers utilize surface modification techniques to address the material's drawbacks in various applications. Techniques such as short peening, heat treatment, laser peening, and laser cladding are common surface modification methods used by researchers. Laser cladding, also known as laser metal deposition (LMD) or laser cladding deposition, is a cutting-edge material processing technique that employs a high-energy laser beam to melt and fuse metallic or composite powders onto a substrate. This technology has found applications in various industries, including manufacturing, aerospace, automotive, and medical, due to its ability to enhance surface properties, repair components, and create complex geometries with precision. During laser cladding, a focused laser beam is utilized to melt the surface of the substrate or a previously deposited layer. The metallic or composite powders are then injected into the molten pool, where they melt and fuse with the substrate, forming a new layer [7, 8]. Commonly used materials include alloys, tool steels, and superalloys. Laser cladding allows for precise control over the heat input, minimizing thermal distortion and ensuring a fine microstructure. Furthermore, the localized nature of the process reduces the impact on the surrounding material, preventing thermal damage to the substrate. It can be employed for various applications, including coating, repairing damaged components, and building up material layer by layer for additive manufacturing.

During laser cladding, several aspects must be taken into consideration, such as ensuring compatibility between the substrate and the clad material, and fine-tuning process parameters, including laser power, scanning speed, and powder flow rate, to achieve optimal results. Variations in these parameters lead to changes in the general shape of the tracks and in the microstructure. Researchers have demonstrated how variations in the cladding parameters affect the microstructure and corrosion resistance of the material. Zhang et al. [9] showed that an increase in laser power increases the grain size of the coating, and decreases the density. Gao et al. [10] investigated the effect of changes in the scanning speed on corrosion resistance. Their results indicate that an increase in the scanning speed leads to a thinner coating thickness, resulting in surface cracks, and decreased corrosion resistance. However, not much research has focused on the effects of changes in cladding parameters on the induced residual stresses of materials.

The stresses that remain in a component after processing and after all externally imposed loads have been eliminated are known as residual stresses [11]. These stresses can result from various processes such as manufacturing, welding, machining, and heat treatment. Understanding residual stresses is crucial because they can significantly affect the mechanical properties and performance of materials. In titanium and titanium alloys, there are numerous sources of residual stresses. Rapid cooling or uneven cooling during manufacturing processes can lead to non-uniform thermal contractions, resulting in residual stresses. Titanium and its alloys may undergo phase transformations during heat treatment, welding, or cooling, leading to volume changes and the development of residual stresses. Mechanical processes such as forging, rolling, or extrusion can induce plastic deformation and generate residual stresses in the material. Moreover, localized heating and cooling during welding can introduce significant residual stresses, particularly near the weld region [12–14]. Residual stresses can influence the mechanical properties of titanium alloys by affecting factors such as fatigue strength, fracture toughness, and corrosion resistance. Residual stresses extend the fatigue life of components by suppressing crack initiation and propagation [15]. Induced residual stresses can be both beneficial and harmful to structural integrity. Wang et al. [16] deliberately induced compressive residual stresses (CRSes) by shot peening to exploit the positive benefits of CRS. However, the presence of tensile residual stresses (TRSes) leads to detrimental effects such as stress corrosion effects and inconsistencies in component precision, which must be avoided.

CRS is crucial for increasing the material's fatigue strength. As the laser beam intensity increases, the surface heats up, causing tension to transition from stretching to compression. Conversely, the remelted zone contracts, and tensile stresses develop due to cooling following laser treatment, potentially leading to the fracture of brittle materials [17]. In this study, the effects of variation in the laser power during the laser cladding process on both the microstructure and induced residual stresses of the cladded CP grade 2 Ti alloy are investigated. The CP grade 2 Ti was coated with the Ti-6Al-4 V alloy, which is an $\alpha + \beta$ alloy characterized by low density and excellent corrosion resistance.

Residual stress can be measured by the X-ray diffraction (XRD) technique. Measuring residual stress using X-ray diffraction (XRD) involves analyzing crystal lattice distortions caused by internal stresses within a material. One commonly used method for this purpose is the $\sin^2\psi$ method, also known as the $\sin^2\psi$ technique or $\sin^2\psi$ analysis. This method utilizes the relationship between the diffraction angle, lattice strain, and residual stress in a crystalline material. The basis of XRD is Bragg's Law, which relates the angle of incidence (θ) of X-rays on a crystal lattice to the interplanar spacing (d) and the wavelength (λ) of the X-rays: $2d\sin\theta = n\lambda$, where n is the order of diffraction. Residual stresses within a material cause lattice strain, leading to distortion in the crystal lattice. The lattice spacing (d) is modified, which affects the diffraction pattern. The lattice strain (ε) is related to the change in the lattice spacing (Δd) by $\varepsilon = \Delta d/d$. The $\sin^2\psi$ method is based on the relationship between the orientation of the crystal lattice planes relative to the incident X-ray beam and the resulting diffraction peak position. The angle ψ is the angle between the normal to the crystal lattice planes and the direction of the incident X-ray beam. The $\sin^2\psi$ term is introduced to account for the arbitrary orientation of the crystal lattice. The stress (σ) can be determined from the lattice strain using Hooke's Law: $\sigma = E\varepsilon$, where E is the elastic modulus of the material. For a polycrystalline material, the relationship between stress, lattice strain, and the XRD diffraction angle (2θ) is given by the $\sin^2\psi$ Eq. (1):

$$sin^2\psi = \frac{1 - cos^2\theta}{1 + cos^2\theta} = \frac{\lambda}{2d}\left(\frac{\partial d}{\partial \theta}\right)^2 + sin^2\theta\frac{E(1+\nu)}{2(1-\nu)}\sigma \tag{1}$$

Previous research has shown the influence of surface modification techniques on the material's microstructure but little on the appearance of residual stresses. Additionally, most researchers have focused on heat treatment as a surface modification technique, with minimal attention given to laser cladding and its influence on the material's residual stresses. Therefore, the aim of this research is to analyze the influence of laser cladding on selected properties of the obtained surface layers for samples made of titanium and titanium alloy Ti-6Al-4V. The laser power was varied and the properties of the resulting samples were compared.

2. Experimental setup

2.1 Materials

The substrate material used in this study is a commercially pure titanium (grade 2) alloy with dimensions of 100 mm × 200 mm × 3 mm. Commercial pure titanium contains 99.0–99.5 wt.% titanium, with the main impurities being iron, carbon, nitrogen, oxygen, and hydrogen [18]. Ti-6Al-4V powder is used as the clad material, and the chemical composition of the powder is summarized in **Table 1** [20].

2.2 Laser cladding

Before cladding, the samples were sandblasted with 200–300 micron glass beads to remove contaminants on the surface, improve surface adhesion, and increase radiation absorption. The laser source used in this study was a Nd:YAG diode laser. A constant laser beam spot size of 2.5 mm was used with a laser power range of 600–1000 W to investigate the effect of the laser power. A total of 15 cladding tracks were made, covering a cladding area of 2700 mm^2, as shown in **Figure 1**. Overlapping coating was achieved with a 50% overlap. The scanning speed, powder flow, and gas flow rate were kept constant at 1.5 m/min, 2.5 rpm, and 3.5 l/min, respectively. These settings were found to be the most efficient for the cladding of titanium alloy [19].

The laser cladding process parameters are summarized in **Table 2**. The specific heat input was calculated using Eq. (1) from [21]. The laser interaction time (which is described as the amount of time the laser is in contact with the material) was kept constant at 0.1 s, which was calculated from Eq. (2) below.

$$H_s = \frac{P}{\nu.D} \tag{2}$$

where H_s is the specific heat input, P is the power in watt, ν is the scanning speed in mm/s, and D is the beam diameter (also known as the spot size).

Element	N_2	H_2	C	O_2	Fe	V	Al	Ti
Wt.%	<0.05	<0.015	<0.08	<0.2	<0.25	3.5–4.5	5.5–6.76	Balance

Table 1.
Chemical composition of Ti6Al4V [19].

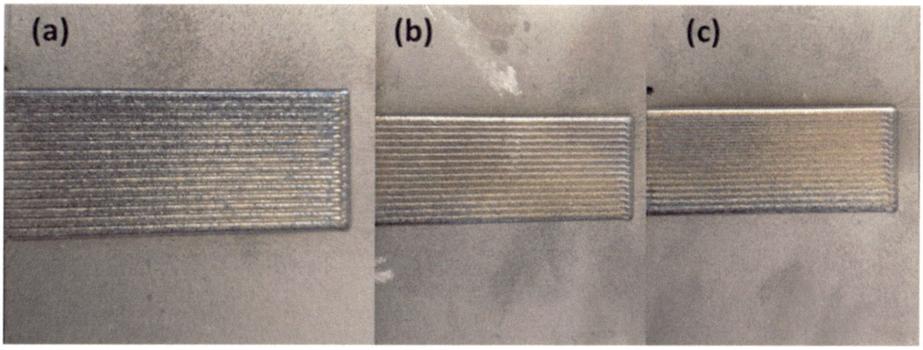

Figure 1.
CP grade 2 titanium after laser cladding. The samples were cladded at laser powers of (a) 600 W, (b) 800 W, and (c) 1000 W.

Sample	Laser power (W)	Spot size (mm)	Scanning speed (m/min)	Powder flow (rpm)	Gas flow (l/min)	Specific heat input (J mm^2)
Ti_{600}	600	2.5	1.5	2.5	3.5	9.6
Ti_{800}	800	2.5	1.5	2.5	3.5	12.8
Ti_{1000}	1000	2.5	1.5	2.5	3.5	16

Table 2.
Laser cladding experimental parameters.

$$R = \frac{D}{\nu} \tag{3}$$

where R is the interaction time in seconds, ν is the scanning speed in mm/s and D is the beam diameter.

2.3 Microstructure analyses

The samples were first cut using the Struers Labotom-5 cutting machine to the desired size. Subsequently, they were then mounted using the Automatic Mounting Press (AMP) 50 at a temperature of 180°C, with a keep time of 580 s, and a cooling temperature of 40°C. Grinding was then performed at a force of 25 N at 300 RPM, using 800, 1200, and 2400 SiC-bonded grinding papers. Polishing was carried out using an MD-Chem polishing disk with a 0.4 μm colloidal silica suspension. The samples were etched in Kroll's reagent for 10 s. The microstructure was determined using a Leica DMI5000M optical microscope (**Figure 2**), with varying the magnifications to obtain clear image of the microstructure.

2.4 Microhardness

The samples used for microstructure analysis were also utilized for microhardness testing. The microhardness of the samples was determined using the Zwick-Roell ZHVμ micro-hardness tester, as depicted in **Figure 3**, with a testing force of 300 gf and a dwell time of 10 s. Hardness tests were performed horizontally across the substrate zone to assess the effects of cladding on the substrate.

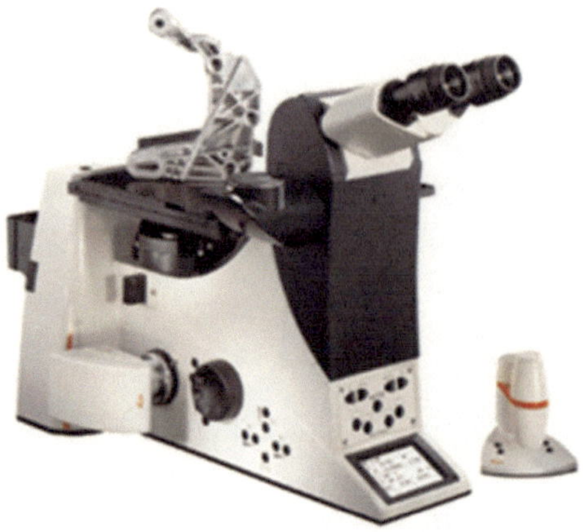

Figure 2.
Leica DMI5000 optical microscope (courtesy of Leica).

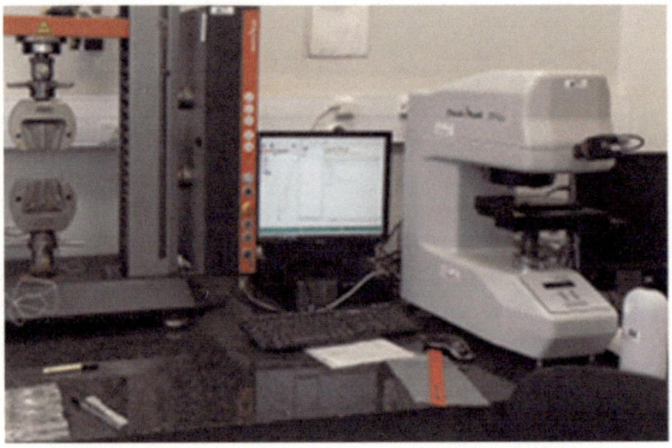

Figure 3.
Zwick-Roell ZHVμ micro-hardness tester.

2.5 Residual stress measurements

The residual stress measurements were carried out using Proto iX-ray diffraction and the sin2ψ method [22]. XRD is a non-destructive technique that allows for stress analysis without altering the material. The sin2ψ method utilizes the planes $d_{\{hkl\}}$ of the crystal lattice to measure the change in d-spacing (Δd). The strain is then determined from Δd, which is subsequently used to calculate the stress. XRD scans are performed at various angles to obtain diffraction peaks. The $\sin^2\psi$ method involves measuring the diffraction peak positions for different orientations of the crystal lattice. The $\sin^2\psi$ equation is then used to fit the experimental data, and the residual stress was calculated by solving for σ in Eq. (4).

$$\sigma_\phi = \frac{E}{(1+\upsilon)\,sin^2\psi}\left[\frac{d_\psi - d_n}{d_n}\right] \tag{4}$$

where σ_ϕ is the stress in the direction of ϕ, E is the Young's modulus, υ is the Poisson's ratio, ψ is the tilt angle, d_ψ is the d-spacing at tilt angle and d_n is the stress-

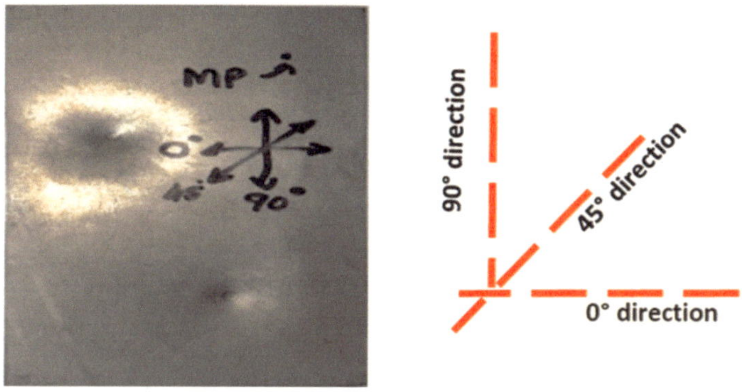

Figure 4.
Tri-axial measurement orientation setup (0°, 45° and 90°) shown on the parent material.

Method	$\sin^2\psi$
X-ray source	Cu_K-Alpha
Wavelength	1.542 Å
Filter	Ni
Plane (hkl)	{213}
$1/2S_2$	11.89×10^{-6} MPa^{-1}
S_1	2.83×10^{-6} MPa^{-1}
Voltage	25 kV
Current	4 mA
d-spacing	0.815
Bragg's angle	142

Table 3.
X-ray diffraction parameters for residual stress measurements.

free d-spacing. A Cu_K-Alpha tube with a wavelength of 1.5418 Å was selected to obtain diffraction peaks from the {213} crystallographic planes of the alpha phase. The instrumental voltage and current were set at 25 kV and 4 mA, respectively. The residual stress measurements were taken from three different angles of rotation (phi (ϕ)) (as shown in **Figure 4**), 0°, 45° and 90° respectively. The Bragg angle was 142°. **Table 3** summarizes the X-ray diffraction parameters used in the stress measurements.

3. Results and discussion

3.1 Laser cladding processing parameters

Several laser cladding parameters must be considered for successful cladding as outlined in Section 3 of this paper. Among these parameters, power and energy density play crucial roles in determining the power transferred to the material and the energy absorbed by it. **Table 4** presents the power density, calculated as the ratio of power to the cross-sectional area of the laser beam, and the energy density, which measures the energy absorbed by the treated material. The power density is determined by the formula $\boldsymbol{I} = \boldsymbol{P}/\boldsymbol{A}$, where I represent intensity in W/cm^2, P is the laser power in W, and A is the cross-section area of the beam [23]. Additionally, the energy density is given by Eq. (5) [24].

$$E = PR/A \tag{5}$$

The power density increases with the laser power, assuming the area of the laser beam remains constant. The power intensity rises by 25% from 600 W to 800 W and by 20% from 800 W to 1000 W. The percentage difference between 600 and 800 W, as well as between 800 and 1000 W, decreases as the laser power increases. The presence of the interaction time (R) in the energy density equation contributes to the main difference compared with power density equation. Both density variables depend on the laser parameters employed during cladding. Results in **Table 4** indicate that the energy density is 10 times less than the power density due to the short interaction time of 0.1 s. The percentage increase in energy density with laser power mirrors to the change observed in the power density values. Overall, the three laser powers under the investigation yielded a good metallurgical bond between the substrate and the cladding material. The influence of the laser power on the mechanical properties of the samples are discussed in the next section of results.

3.2 Microstructure

Figure 5 presents the results obtained through optical microscopic examination at 10× magnification. The microstructure of the substrate (CP Ti grade 2) is shown in **Figure 5(a)** revealing the dominance of the hexagonal α phase on the CP Ti, as expected, with small α grains. The small α grains result from temperature adjustments (typically around 700–750°C) during the recrystallization treatment when forming CP Ti [24]. **Figure 5(b)–(d)** illustrate the microstructure of the laser-clad samples, where the laser-clad coatings are divided into three zones: the clad zone (CZ), the interface zone (IZ), and the heat-affected zone (HAZ). A robust metallurgical bond is observed between the clad zone and the substrate base material.

Figure 5 shows that the crystal morphology of the cladded samples changed dramatically compared with the initial microstructure due to solid phase transformation. At lower laser power (600 W), as shown in **Figure 5(b)**, the clad layer appears to have a uniform distribution of particles and is free of pores. However, gas pores and cracks are noticeable, especially in the IZ and CZ, at laser powers of 800 and 1000 W (**Figure 5(c)** and **(d)**). This may be attributed to the high laser power, which results in excessive heat absorption, leading to material evaporation. This issue can be addressed by adjusting the power flow rate with laser power [25]. Moreover, the HAZ consist of

Laser power (W)	Power density (W/cm^2)	Energy density (J/cm^2)	Power, energy density increase (%)
600	12223.1	1222.3	—
800	16297.5	1629.7	25
1000	20371.8	2037.2	20

Table 4.
Laser processing energy parameters.

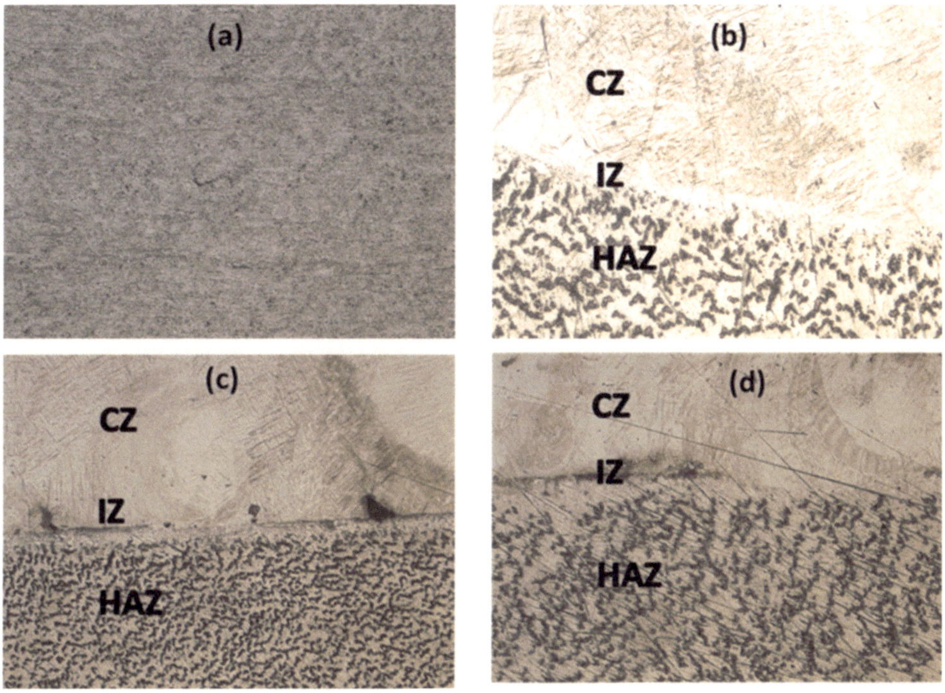

Figure 5.
Microstructures of (a) parent Ti, (b) Ti_{600} sample, (c) Ti_{800}, (d) Ti_{1000} with 10× magnification. (b)–(d) show the junction of the clad area and the substrate, clad zone (CZ), interface zone (IZ), and substrate zone (SZ).

a dual-phase microstructure with α grains (represented by light color) and intergranular β grains (in black color). The number and density of intergranular β grains increases with increasing laser power. According to [26], the microstructure for HAZ ranged from fine to coarse globular alpha structure as the laser power is increased. However, unlike in this current where pure titanium was used as a substrate, they used Ti alloy as the substrate.

Furthermore, the martensitic transformation is clearly evident in the clad samples compared with the parent sample. The martensitic transformation results in the formation of a hexagonal martensite crystal structure (α′) with lath martensite. Studies by [26, 27] show that increasing laser power lead to a deeper melt, which result in much thicker martensite laths. In **Figure 5**, the same phenomenon is observed, where Ti_{600} indicate much fine martensite than the Ti_{800} and Ti_{1000}, respectively.

3.3 Microhardness

During the cladding process, all parameters remained constant except for the laser power, which varied to assess its impact on the microhardness of the clad material. It is noteworthy that the microhardness of the clad samples was measured in the substrate zone, rather than the clad zone, to evaluate the influence of cladding on the substrate. The microhardness on the substrate was reduced by up to 11.6% as a result of cladding process when compared to the parent material. According to [28], the microhardness of the substrate is 30.4% less than that of the coating due to difference in the grain size of the microstructure. The average microhardness of the clad samples is summarized in **Table 5** and compared with that of the parent sample. Overall, the

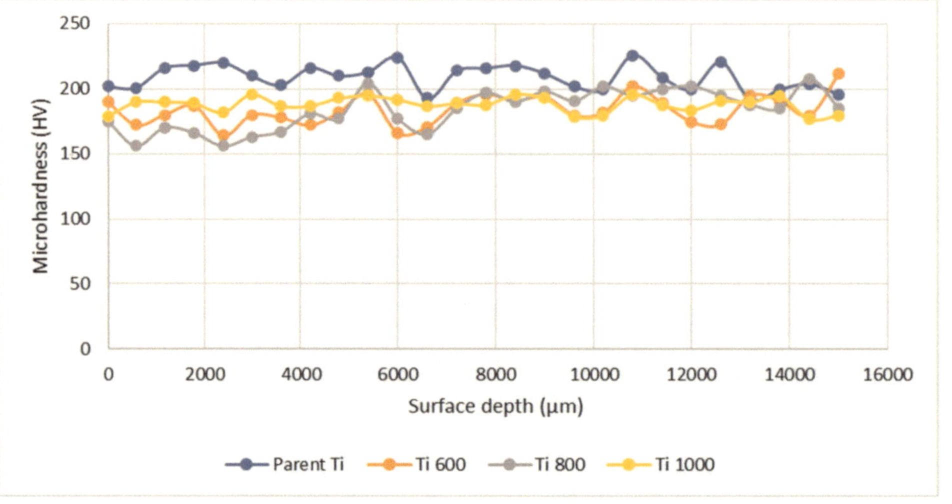

Figure 6.
Microhardness profiles of the cladded samples (with different laser power) versus the depth of the surface.

microhardness of the parent material was higher than that of the clad samples, primarily due to the fact that the microhardness were measured in the substrate zone instead of the clad zone.

Figure 6 illustrates the microhardness depth profile of the clad samples and the parent material. The microhardness distribution across the surface depth generally increases with an increasing depth. The parent material consistently exhibits higher microhardness values across all depth levels, whereas the clad samples display fluctuations with increasing depth values. The microhardness distribution fluctuates across the surface depth, with the fluctuations ranging between 1 and 52 HV. Ti_{1000} fluctuates less among the three cladded samples, with the highest fluctuation of up to 19 HV across the depth. Ti_{600} and Ti_{800} have their highest fluctuations up to 48 and 52 HV, respectively.

Table 5 presents the microhardness results for the different samples. Analysis of **Table 5** shows that the parent material has a higher microhardness value than the cladded samples, this is a result of measuring the microhardness on the substrate instead of the clad zone. The difference in the microhardness between the parent material and the cladded sample ranges between 9.4% and 11.6. **Figure 7** illustrates a non-linear relationship between microhardness and power density which is equivalent to laser power %. Microhardness decreases by 2.2% from 600 to 800 W and then increases by 2.4% from 800 to 1000 W. Similar and contrasting results are observed in literature, Abdulrahman et al. [29] demonstrated non-linear relationship between microhardness and laser power, showing a fluctuation of the microhardness as the laser power is increased, attributed to the slow cooling during solidification. In contrast, according to [26] the average microhardness increases with an increase in laser

Parent Ti	Ti_{600}	Ti_{800}	Ti_{1000}
207.8	187.8	183.7	188.2

Table 5.
Average Vickers microhardness (HV) for the cladded Ti samples.

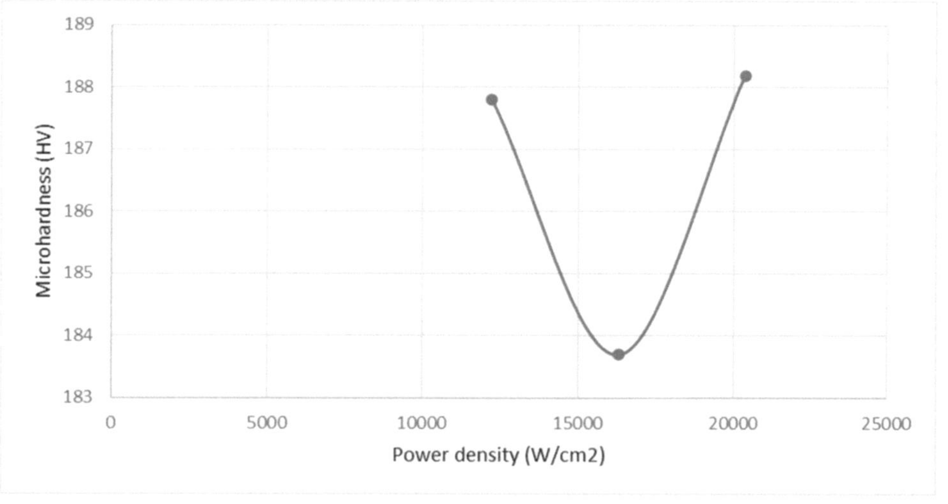

Figure 7.
Graph of microhardness versus power density of the radiated beam.

power due to changes in microstructure. It was noted that the microstructure range between fine and thick martensite structure as the laser power is increased. Conversely [30], stated that the average microhardness decrease with increasing laser power due to differences in microstructure activated by lengthening cooling speed of the alloy layer under a higher laser power.

Overall, the maximum microhardness value was observed at 1000 W. This could be attributed to the more intense and rapid heating of the material being deposited at higher laser power, resulting in a faster cooling rate. Another possibility is that the increase in laser power leads to higher energy input into the material, resulting in more complete melting of the deposited material. Consequently, during solidification, smaller and more uniform grains are formed, contributing to higher microhardness.

3.4 Residual stress

The residual stress measurements were conducted using XRD and $\sin^2\psi$ method. Measurements were taken for both the parent material (CP Ti grade 2 alloy) and the cladded samples at three different rotation angles, as depicted in **Figure 4**, with results summarized in **Table 6**. Analysis of **Table 6**, reveals that the investigated samples exhibit TRS, with the highest residual stress observed at 600 W. Specifically, Ti_{600} exhibit the maximum residual stress, averaging 215.41 ± 31.36 MPa, followed by Ti_{800} with 157.03 ± 24.93 MPa, and Ti_{1000} with an average residual stress of 135.14 ± 25.22 MPa, respectively. In contrast, the parent material displayed the lowest residual stress of 44.99 ± 27.19 MPa, indicating a significant increase in the induced residual stress due to the cladding process of the Ti alloy. Furthermore, as depicted in **Figure 8**, an increase in the laser power correlates with a decrease in the residual stress. Specifically, cladding the parent material with a laser power of 600 W resulted in a significant average increase in TRS of 478.8%. Similarly, when the laser power was increased to 1000 W, the average TRS increased by 300%. According to [31], modification of the pure titanium by laser treatment with average power of 25 W increased the stresses by 225%. Simulation studies shows non-linear relationship

0°	45°	90°	Average
212.59 ± 36.43	245.92 ± 24.16	187.71 ± 33.48	215.41 ± 31.36
179.19 ± 30.93	141.09 ± 21.83	150.82 ± 22.03	157.03 ± 24.93
149.63 ± 31.48	127.73 ± 22.63	128.07 ± 21.56	135.14 ± 25.22
35.07 ± 27.28	40.44 ± 28.93	59.45 ± 25.35	44.99 ± 27.19

Table 6.
Residual stress measurements (in MPa) for cladded samples at different rotation angles.

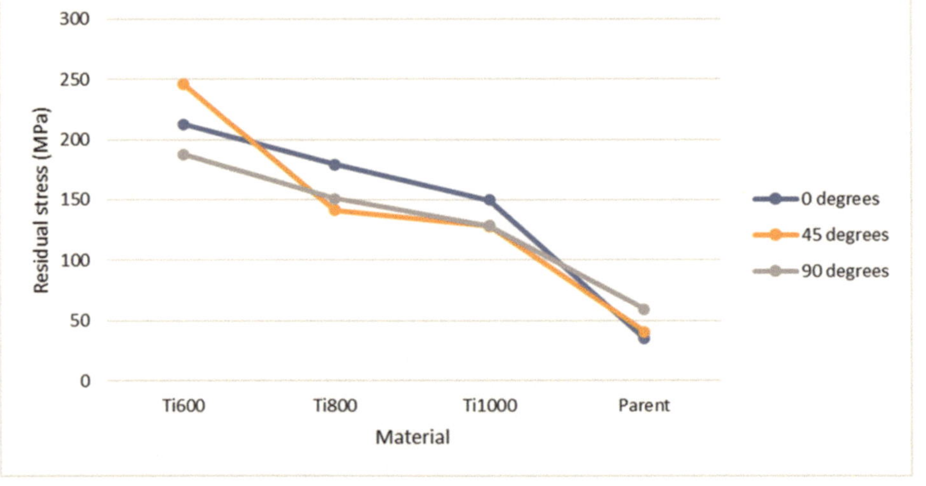

Figure 8.
Residual stress measurements for different materials.

between laser power and residual stress, however, the trend is not consistent with different scanning speeds [32].

Figure 9 illustrates that the residual stress does not follow a clear trend when the rotation angle is altered. The maximum induced residual stress occurs at 45° for a lower laser power of 600 W, at 0° for 800 W and 1000 W, and at 90° for the parent material. This observation aligns with findings by Lai et al. [33], who also noted the absence of a defined trend between the scanning angle and residual stress.

Compressive and tensile residual stresses in materials can exert opposing effects on the material's properties and performance. CRS can benefit fatigue-loaded structures by delaying crack growth, thereby enhancing the material's fatigue life [34]. Conversely, TRS decrease the fatigue life of the material as they can initiate crack growth and propagation. It would be insightful to investigate the effects of laser cladding processing conditions on induced residual stress and its correlation with the material's microstructure. Microstructure control is crucial for maintaining a balanced relationship between tensile and compressive residual stresses [35].

Considering the obtained results for the cladded samples and the effects of laser power, it can be inferred that an increase in laser power leads to a reduction in the TRS, an increase in hardness, and a martensitic α' microstructure. The formation of α' martensitic promotes high tensile strength but results in low ductility of the alloy [36]. Therefore, the difference in the microstructure, can account for the variation in the induced residual stresses. Furthermore, it is important to note that the residual

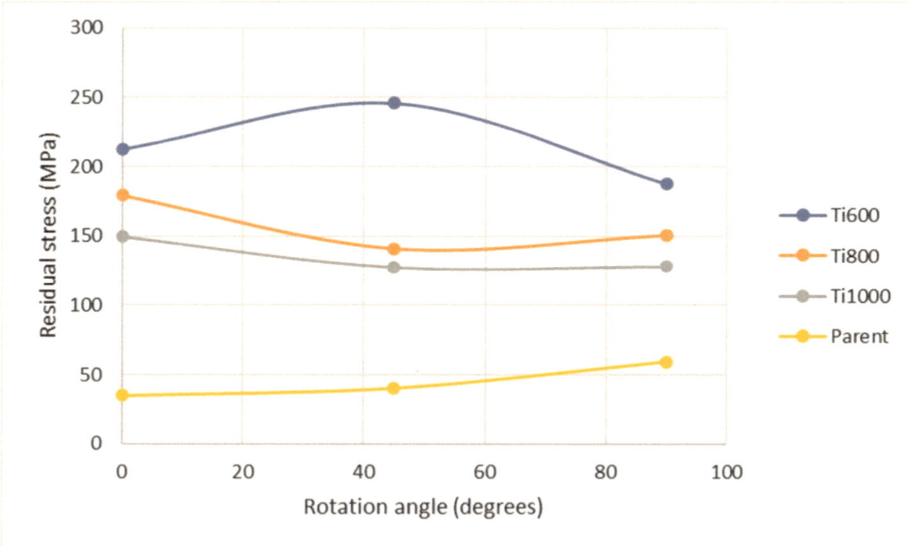

Figure 9.
Relationship between residual stress and rotation angle for different materials.

stresses in this investigation were only measured on the surface, which is a limitation because the residual stress also depends on the material's depth.

The hardness of the material is determined by measuring the amount of persistent deformation caused by a set force exerted on the penetrator, defining its resistance to permanent or plastic deformation. The yield strength, elastic modulus, and strain-hardening properties of a metal primarily determine its resistance to deformation [37]. There is a growing interest among scientists and engineers in establishing the relationship between a material's hardness and induced residual stress. The current study focused on the effects of laser power, revealing a higher hardness value at 1000 W. The Ti_{1000} sample also indicates the lowest TRS, which approaches CRS. By enhancing surface hardening and inducing higher CRS, deeper crack initiation sites and superior fatigue life can be achieved [38]. However, the Ti_{800} sample displays inconsistent results with this trend. It exhibits the lowest hardness value among the three cladded samples while having a TRS value lower than Ti_{600}.

Overall, the results are consistent with findings by [31]. Jażdżewska et al. [31], who studied the mechanical properties and residual stress measurements of grade 4 Ti and Ti alloys after laser treatment. Their results indicated an increase in the tensile stresses after laser modification, along with an increase in material hardness. Considering both the current study and that of Jażdżewska, it can be concluded that increasing the CRS or, decreasing TRS as observed in the current study, improves the material hardness, thereby enhancing fatigue-life and crack initiation resistance. Moreover, it enhances corrosion and abrasion resistance. However, the current study suggests that processing parameters significantly impact outcomes, as indicated by the deviation of Ti_{800}.

4. Conclusions

1. The successful laser cladding of Ti6Al4V alloy onto pure titanium alloy resulted in a significant transformation in the microstructure compared to the initial state

of the parent material. The microstructure of the cladding zone exhibited a prominent martensite lath structure, with the thickness of martensite laths increasing proportionally with the laser power.

2. Analysis of surface microhardness revealed a nonlinear correlation with laser power. The microhardness distribution across the substrate depth demonstrated a general increase with depth, with higher laser power exhibiting reduced fluctuations in microhardness distribution compared to lower power settings.

3. Comparative analysis of residual stresses between the cladded samples and the parent material revealed a notable increase in average residual stresses on the surface of the cladded samples, ranging from 300 to 478.8% depending on the laser power. These residual stresses, predominantly tensile, signify the induction of stress during the laser cladding process. Furthermore, there was a discernible decrease in residual stresses with increasing laser power, underscoring the influence of processing parameters on induced stress levels and suggesting potential benefits in controlling residual stresses during material processing.

Acknowledgements

Walter Sisulu University, Department of Research and Innovation.
Walter Sisulu University, Department of Mechanical Engineering.
Council for Scientific & Industrial Research (CSIR) Photonics Centre, Pretoria, South Africa.
Nelson Mandela University eNtsa Laboratory, Qheberha, South Africa.

Conflict of interest

The authors declare no conflict of interest.

Author details

Tankiso Lawrence Ngake* and Kadephi Vuyolwethu Mjali
Department of Mechanical Engineering, Walter Sisulu University, Butterworth, Eastern Cape, South Africa

*Address all correspondence to: tngake@wsu.ac.za

References

[1] Wang K. The use of titanium for medical applications in the USA. Materials Science and Engineering. 1996; **A213**:134-137. DOI: 10.1016/0921-5093(96)10243-4

[2] Boyer RR. Titanium for aerospace: Rationale and applications. Advanced Performance Materials. 1995;**2**:349-368. DOI: 10.1007/BF00705316

[3] Boyer R. ASM Handbook Metallography and Microstructure. 9th ed. USA: ASM International; 2000. p. 460

[4] Rae W, Rahimi S. Evolution of microstructure and residual stress in hot rolled Ti-6Al-4V plates subjected to different heat treatment conditions. In: Conference: European Conference on Residual Stresses 10 (ECRS10). Materials Research Proceedings. Vol. 6. USA: Materials Research Forum LLC; 2018. pp. 171-176. DOI: 10.21741/9781945291890-27

[5] Deepak JR, Joy N, Krishnamoorthy A, Jaswanth CP, Harish G. Gas nitriding of CP grade – 2 commercially pure titanium and Ti6Al4V grade – 5 titanium alloy. Materials Today: Proceedings. 2021;**44** (Part 5):3744-3750. DOI: 10.1016/j.matpr.2020.11.586

[6] Umapathi A, Swaroop S. Measurement of residual stresses in titanium alloys using synchrotron radiation. Measurement. 2019;**140**: 518-525. DOI: 10.1016/j.measurement.2019.04.021

[7] Meriaudeau F, Truchetet F, Grevey D, Vannes AB. Laser cladding process and image processing. Journal of Lasers in Engineering. 1997;**6**:161-187. DOI: 10.1117/12.251167

[8] Vilar R. Laser cladding. Journal of Laser Applications. 1999;**11**:64-79. DOI: 10.2351/1.521888

[9] Zhang H, Pan Y, Zhang Y, Lian G, Cao Q, Yang J. Influence of laser power on the microstructure and properties of in-situ NbC/WCoB–TiC coating by laser cladding. Materials Chemistry and Physics. 2022;**290**:126636. DOI: 10.1016/j.matchemphys.2022.126636

[10] Gao W, Wang S, Hu K, Jiang X, Yu H, Sun D. Effect of laser cladding speed on microstructure and properties of titanium alloy coating on low carbon steel. Surface and Coating Technology. 2022;**451**:129029. DOI: 10.1016/j.surfcoat.2022.129029

[11] Ulutan D, Ozel T. Machining induced surface integrity in titanium and nickel alloys: A review. International Journal of Machine Tools and Manufacture. 2011;**51**:250-280. DOI: 10.1016/j.ijmachtools.2010.11.003

[12] Parlevliet PP, Bersee HEN, Beukers A. Residual stresses in thermoplastic composites—A study of the literature—Part I: Formation of residual stresses. Composites Part A: Applied Science and Manufacturing. 2006;**37**(11):1847-1857. DOI: 10.1016/j.compositesa.2005.12.025

[13] Chen S-G, Zhang Y-D, Wu Q, Gao H-J, Gao Z-H, Li X. Effect of solid-state phase transformation on residual stress of selective laser melting Ti6Al4V. Materials Science and Engineering: A. 2021;**819**:141299. DOI: 10.1016/j.msea.2021.141299

[14] Mehdi B, Badji R, Ji V, Allili B, Bradai D, Deschaux-Beaume F, et al. Microstructure and residual stresses in Ti-6Al-4V alloy pulsed and unpulsed TIG welds. Journal of Materials Processing Technology. 2016;**231**: 441-448. DOI: 10.1016/j.jmatprotec.2016.01.018

[15] Nalla R, Altenberger I, Noster U, Liu G, Scholtes, Ritchie BR. On the influence of mechanical surface treatments—Deep rolling and laser shock peening—On the fatigue behavior of Ti–6Al–4V at ambient and elevated temperatures. Materials Science and Engineering A. 2003;**355**:216-230. DOI: 10.1016/S0921-5093(03)00069-8

[16] Wang S, Li Y, Yao M, Wang R. Compressive residual stress introduced by shot peening. Journal of Materials Processing Technology. 1998;**73**:64-73. DOI: 10.1016/S0924-0136(97)00213-6

[17] Jażdżewska M. Effects of CO2 and Nd:YAG laser remelting of the Ti6Al4V alloy on the surface quality and residual stresses. Advances in Materials Science. 2020;**20**:82-90. DOI: 10.2478/adms-2020-0005

[18] Ion JC. Laser Processing of Engineering Materials, Principles, Procedure and Industrial Application. Oxford, UK: Elsevier Butterworth Heinemann; 2005

[19] du Plooy R, Akinlabi E. T: Analysis of laser cladding of titanium alloy. Materials Today: Proceedings. 2018;**5**: 19594-19603. DOI: 10.1016/j.matpr.2018.06.322

[20] Azom: Titanium Alloys - Ti6Al4V Grade 5. 2002. [Online]. Available from: http://www.azom.com/article.aspx?ArticleID=1547

[21] Zhechao F, Hongwei F. Study on selective laser melting and heat treatment of Ti-6Al-4V alloy. Results in Physics. 2018;**10**:660-664. DOI: 10.1016/j.rinp.2018.07.008

[22] Hosseini S, Farajollahi M, Ebrahimi M. Residual stress, fatigue behavior, and mechanical properties of equal-channel angular pressed commercial pure titanium. Journal of Materials Research and Technology. 2024;**28**:3297-3305. DOI: 10.1016/j.jmrt.2023.12.265

[23] Matys J, Dominiak M, Flieger R. Energy and power density: A key factor in lasers studies. Journal of Clinical and Diagnostic Research. 2015;**9**(12):ZL01-ZL02. DOI: 10.7860/JCDR/2015/15561.6955

[24] Trushin ES, Dasaev MR, Kalakutskaya OV, Voloshenko AP. On the method of calculating the energy density of laser radiation in a metal surface modification. Journal of Physics: Conference Series. 2021;**2124**:012014. DOI: 10.1088/1742-6596/2124/1/012014

[25] Ali SR, Hussein AHA, Nofal A, Elnaby SIH, Elgazzar H. A contribution to laser cladding of Ti-6Al-4V titanium alloy. Metallurgical Research & Technology. 2019;**116**:634. DOI: 10.1051/metal/2019060

[26] Mahamood RM, Akinlabi ET, Shukla M, Pityana S. Laser metal deposition of Ti6Al4V: A study on the effect of laser power on microstructure and microhardness. In: Proceedings of the International MultiConference of Engineers and Computer Scientists 2013 Vol II. Hong Kong: IMECS 2013; March 13-15, 2013; Hong Kong

[27] Cottamac R, Brandt M. Laser cladding of Ti-6Al-4V powder on Ti-6Al-4V substrate: Effect of laser cladding parameters on microstructure. Physics Procedia. 2011;**12**:323-329. DOI: 10.1016/j.phpro.2011.03.041

[28] Wang C, Li J, Wang T, Chai L, Deng C, Wang Y, et al. Microstructure and properties of pure titanium coating on Ti-6Al-4V alloy by laser cladding. Surface & Coatings Technology. 2021;

416:127137. DOI: 10.1016/j.surfcoat.2021.127137

[29] Abdulrahman KO, Mahamood RM, Akinlabi ET, Adediran AA. Effect of laser power on the microstructure and mechanical properties of laser deposited titanium aluminide composite. Advances in Materials and Processing Technologies. 2022;**8**(Suppl 3): 1305-1316. DOI: 10.1080/2374068X.2021.1945268

[30] Lee K-H, Choi S-W, Suh J, Kang C-Y. Effect of laser power and powder feeding on the microstructure of laser surface alloying hardened H13 steel using SKH51 powder. Materials and Design. 2016;**95**:173-182. DOI: 10.1016/j.matdes.2016.01.079

[31] Jażdżewska M, Kwidzińska DB, Seyda W, Fydrych D, Andrzej Zieliński A. Mechanical properties and residual stress measurements of grade IV titanium and Ti-6Al-4V and Ti-13Nb-13Zr titanium alloys after laser treatment. Materials. 2021;**14**:6316. DOI: 10.3390/ma14216316

[32] Xiao Z, Chen C, Zhu H, Hu Z, Nagarajan B, Guo L, et al. Study of residual stress in selective laser melting of Ti6Al4V. Materials and Design. 2020; **193**:108846. DOI: 10.1016/j.matdes.2020.108846

[33] Lai Y, Liu W, Zhao J, Zhao Y, Wang F, Han W. Experimental study on residual stress in titanium alloy laser additive manufacturing. Applied Mechanics and Materials. 2013;**431**: 20-26. DOI: 10.4028/www.scientific.net/AMM.431.20

[34] Neto DM, Borges MF, Sérgio ER, Antunes FV. Effect of residual stresses on fatigue crack growth: A numerical study based on cumulative plastic strain at the crack tip. Materials. 2022;**15**:2156. DOI: 10.3390/ma15062156

[35] Edwards P, Ramulu M. Surface residual stresses in Ti-6Al-4V friction stir welds: Pre- and post-thermal stress relief. Journal of Materials Engineering and Performance. 2015;**24**:3263-3270. DOI: 10.1007/s11665-015-1610-2

[36] Kazantseva N, Krakhmalev P, Thuvander M, Yadroitsev I, Vinogradova N, Ezhov I. Martensitic transformations in Ti-6Al-4V (ELI) alloy manufactured by 3D printing. Materials Characterization. 2018;**146**:101-112. DOI: 10.1016/j.matchar.2018.09.042

[37] Frankel J, Abbate A, and Scholz W: Measurement and theory of the dependence of hardness on residua stress. Journal of Experimental Mechanics. June 1993:164-168

[38] Maleki E, Bagherifard S, Guagliano M. Correlation of residual stress, hardness and surface roughness with crack initiation and fatigue strength of surface treated additive manufactured AlSi10Mg: Experimental and machine learning approaches. Journal of Materials Research and Technology. 2023;**24**: 3265-3283. DOI: 10.1016/j.jmrt.2023.03.193